Methods in Molecular Biology

For further volumes:
http://www.springer.com/series/7651

For over 35 years, biological scientists have come to rely on the research protocols and methodologies in the critically acclaimed *Methods in Molecular Biology* series. The series was the first to introduce the step-by-step protocols approach that has become the standard in all biomedical protocol publishing. Each protocol is provided in readily-reproducible step-by-step fashion, opening with an introductory overview, a list of the materials and reagents needed to complete the experiment, and followed by a detailed procedure that is supported with a helpful notes section offering tips and tricks of the trade as well as troubleshooting advice. These hallmark features were introduced by series editor Dr. John Walker and constitute the key ingredient in each and every volume of the *Methods in Molecular Biology* series. Tested and trusted, comprehensive and reliable, all protocols from the series are indexed in PubMed.

Bio-Carrier Vectors

Methods and Protocols

Edited by

Kumaran Narayanan

School of Science, Monash University Malaysia, Bandar Sunway, Selangor, Malaysia

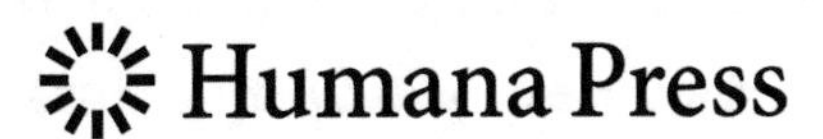

Editor
Kumaran Narayanan
School of Science
Monash University Malaysia
Bandar Sunway, Selangor, Malaysia

ISSN 1064-3745 ISSN 1940-6029 (electronic)
Methods in Molecular Biology
ISBN 978-1-0716-0945-3 ISBN 978-1-0716-0943-9 (eBook)
https://doi.org/10.1007/978-1-0716-0943-9

Cover Illustration Caption: Scanning Electron microscopy of fibrosarcoma (HT1080) cells being invaded by engineered E. coli (blue).

This Humana imprint is published by the registered company Springer Science+Business Media, LLC, part of Springer Nature.
The registered company address is: 1 New York Plaza, New York, NY 10004, U.S.A.

Preface

Through decades of developing gene and drug delivery technologies, we now better appreciate that the traditional approach depending on a single vector system may not be ideal. Although every vector class has its own demonstrated strengths, individually they all have limitations—low delivery efficiency, safety concerns, poor persistence, lack of targeting, or insufficient regulatable control—that one way or another constrain their effectiveness. This realization has led to efforts to adopt and combine the best capabilities from natural and artificial vector systems to assemble improved delivery technologies aimed for specific applications.

Bio-Carrier Vectors looks at delivery systems from this perspective. It discusses vectors other than the traditional viral and non-viral systems and explores ideas for blending the best features of bacteria (Chaps. 1–3), nanoparticles (Chaps. 4–6), peptides (Chaps. 7–10), and hybrid systems (Chaps. 11–15) for the delivery of biomaterials into cells. The final chapter (Chap. 16) discusses application of shotgun proteomics and mass spectrometry as a tool to analyze the proteomic profile changes in cells that result from these interventions.

Each chapter in *Bio-Carrier Vectors* starts with an Introduction that provides an overview of the technology and is followed by a Materials section that describes the required reagents and materials that need to be prepared before starting the experiment. Next comes the Methods section that is formatted as a step-by-step protocol that could be used directly on the bench as the researcher carries out the experiment. At the end of the chapter is the Notes section, which is a distinct feature found in the *Methods in Molecular Biology* series of books. Each Note provides timely advice and guidance from experts that aim to help even first-time users of the method to navigate the steps and compete the experiments successfully. These could include unpublished pointers and suggestions, usually not found in the materials and methods section of scientific publications, which could foster a more nuanced understanding of the protocol.

I would like to especially thank John M. Walker, the Series Editor, for his expert editing guidance and for his occasional nudging that kept my momentum going. My appreciation goes to all the authors who have gifted their time and effort to share their protocols that will no doubt contribute to the work of many researchers who use these techniques. At Springer, Anna Rakovsky provided excellent support during the final stages of editing and guided its finalization to print. And most of all, I am thankful to my wife Stella Salleh for her support and understanding throughout this project that helped bring it to completion.

Bandar Sunway, Malaysia *Kumaran Narayanan*

Contents

Contributors

SYAFIQ ASNAWI ZAINAL ABIDIN • *Jeffrey Cheah School of Medicine and Health Sciences, Monash University Malaysia, Jalan Lagoon Selatan, Bandar Sunway, Selangor Darul Ehsan, Malaysia*

ALESSANDRO CASNATI • *Department of Chemistry, Life Sciences and Environmental Sustainability, Parma University, Parma, Italy*

SING YIAN CHEW • *School of Chemical and Biomedical Engineering, Nanyang Technological University, Singapore, Singapore; Lee Kong Chian School of Medicine, Nanyang Technological University, Singapore, Singapore*

JIAH SHIN CHIN • *School of Chemical and Biomedical Engineering, Nanyang Technological University, Singapore, Singapore; NTU Institute of Health Technologies, Interdisciplinary Graduate School, Nanyang Technological University, Singapore, Singapore*

SHIOW-HER CHIOU • *Graduate Institute of Microbiology and Public Health, National Chung Hsing University, Taichung, Taiwan*

WAI HON CHOOI • *School of Chemical and Biomedical Engineering, Nanyang Technological University, Singapore, Singapore*

ROBERTO CORRADINI • *Department of Chemistry, Life Sciences and Environmental Sustainability, Parma University, Parma, Italy*

BERTRAND CZARNY • *School of Materials Science and Engineering, Nanyang Technological University, Singapore, Singapore; Lee Kong Chian School of Medicine, Nanyang Technological University, Singapore, Singapore*

MOUMITA DEY • *Department of Natural Resources and Environmental Studies, National Dong Hwa University, Hualien, Taiwan*

KYLE S. FELDMAN • *Department of Developmental Biology, Rangos Research Center, University of Pittsburgh School of Medicine, Pittsburgh, PA, USA*

ALESSIA FINOTTI • *Section of Biochemistry and Molecular Biology, Department of Life Sciences and Biotechnology, Ferrara University, Ferrara, Italy*

ROBERTO GAMBARI • *Section of Biochemistry and Molecular Biology, Department of Life Sciences and Biotechnology, Ferrara University, Ferrara, Italy; Interuniversity Consortium for Biotechnology, Trieste University, Trieste, Italy*

JESSICA GASPARELLO • *Section of Biochemistry and Molecular Biology, Department of Life Sciences and Biotechnology, Ferrara University, Ferrara, Italy*

WEI JIANG GOH • *Department of Pharmacy, National University of Singapore, Singapore, Singapore; NUS Graduate School for Integrative Sciences and Engineering, Centre for Life Sciences (CeLS), Singapore, Singapore*

MAŁGORZATA GRABOWSKA • *Department of Molecular Neurooncology, Institute of Bioorganic Chemistry Polish Academy of Sciences, Poznan, Poland*

BARTOSZ F. GRZEŚKOWIAK • *NanoBioMedical Centre, Adam Mickiewicz University in Poznan, Poznań, Poland*

MICHELLE CLAIRE GUGLER • *School of Science, Monash University Malaysia, Bandar Sunway, Selangor Darul Ehsan, Malaysia*

DAN HE • *Chongqing Research Center for Pharmaceutical Engineering, Chongqing Medical University, Chongqing, China*

ANDREW HILL • *Department of Chemical and Biological Engineering, University at Buffalo, The State University of New York, Buffalo, NY, USA*
NATALIE J. HOLL • *Department of Biological Sciences, Missouri University of Science and Technology, Rolla, MO, USA*
YUE-WERN HUANG • *Department of Biological Sciences, Missouri University of Science and Technology, Rolla, MO, USA*
TAKU KAITSUKA • *Faculty of Life Sciences, Department of Molecular Physiology, Kumamoto University, Kumamoto, Japan*
HAN-JUNG LEE • *Department of Natural Resources and Environmental Studies, National Dong Hwa University, Hualien, Taiwan*
RADOSŁAW MRÓWCZYŃSKI • *NanoBioMedical Centre, Adam Mickiewicz University in Poznan, Poznań, Poland*
RAKESH NAIDU • *Jeffrey Cheah School of Medicine and Health Sciences, Monash University Malaysia, Jalan Lagoon Selatan, Bandar Sunway, Selangor Darul Ehsan, Malaysia*
KUMARAN NARAYANAN • *School of Science, Monash University Malaysia, Bandar Sunway, Selangor, Malaysia*
ALLAN WEE REN NG • *School of Science, Monash University Malaysia, Bandar Sunway, Selangor Darul Ehsan, Malaysia*
ANDREW N. OSAHOR • *School of Science, Monash University Malaysia, Bandar Sunway, Selangor Darul Ehsan, Malaysia*
IEKHSAN OTHMAN • *Jeffrey Cheah School of Medicine and Health Sciences, Monash University Malaysia, Jalan Lagoon Selatan, Bandar Sunway, Selangor Darul Ehsan, Malaysia*
YI-HSUAN OU • *Department of Pharmacy, National University of Singapore, Singapore, Singapore*
DONGWON PARK • *Department of Chemical and Biological Engineering, University at Buffalo, The State University of New York, Buffalo, NY, USA*
GIORGIA PASTORIN • *Department of Pharmacy, National University of Singapore, Singapore, Singapore; NUS Graduate School for Integrative Sciences and Engineering, Centre for Life Sciences (CeLS), Singapore, Singapore*
MARIA P. PAVLOU • *Dualsystems Biotech AG, Schlieren, Switzerland*
BLAINE A. PFEIFER • *Department of Chemical and Biological Engineering, University at Buffalo, The State University of New York, Buffalo, NY, USA*
JANARTHANAN PUSHPAMALAR • *School of Science, Monash University Malaysia, Bandar Sunway, Selangor Darul Ehsan, Malaysia; Monash-Industry Palm Oil Education and Research Platform (MIPO), Monash University Malaysia, Bandar Sunway, Selangor Darul Ehsan, Malaysia*
KATARZYNA ROLLE • *Department of Molecular Neurooncology, Institute of Bioorganic Chemistry Polish Academy of Sciences, Poznan, Poland; Centre for Advanced Technologies, Poznan, Poland*
FRANCESCO SANSONE • *Department of Chemistry, Life Sciences and Environmental Sustainability, Parma University, Parma, Italy*
THENAPAKIAM SATHASIVAM • *Monash-Industry Palm Oil Education and Research Platform (MIPO), Monash University Malaysia, Bandar Sunway, Selangor Darul Ehsan, Malaysia*
MOGANAVELLI SINGH • *Nano-Gene and Drug Delivery Group, Discipline of Biochemistry, School of Life Sciences, University of KwaZulu-Natal, Durban, South Africa*

NONTAPHAT THONGSIN • *Siriraj Center for Regenerative Medicine, Research Department, Faculty of Medicine Siriraj Hospital, Mahidol University, Bangkok, Thailand; Department of Immunology, Faculty of Medicine Siriraj Hospital, Mahidol University, Bangkok, Thailand*
KAZUHITO TOMIZAWA • *Faculty of Life Sciences, Department of Molecular Physiology, Kumamoto University, Kumamoto, Japan*
MATTHIAS WACKER • *Department of Pharmacy, National University of Singapore, Singapore, Singapore*
JIONG-WEI WANG • *Department of Surgery, Yong Loo Lin School of Medicine, National University of Singapore, Singapore, Singapore; Cardiovascular Research Institute, National University Heart Centre, Singapore, Singapore*
TINGTING WANG • *Biochemistry and Molecular Biology Laboratory, Experimental Teaching and Management Center, Chongqing Medical University, Chongqing, China*
METHICHIT WATTANAPANITCH • *Siriraj Center for Regenerative Medicine, Research Department, Faculty of Medicine Siriraj Hospital, Mahidol University, Bangkok, Thailand*
YAN WU • *Chongqing Research Center for Pharmaceutical Engineering, Chongqing Medical University, Chongqing, China*
XUEMEI XIE • *Chongqing Research Center for Pharmaceutical Engineering, Chongqing Medical University, Chongqing, China*
MALIHA ZAHID • *Department of Developmental Biology, Rangos Research Center, University of Pittsburgh School of Medicine, Pittsburgh, PA, USA*
JINGQING ZHANG • *Chongqing Research Center for Pharmaceutical Engineering, Chongqing Medical University, Chongqing, China*
CAILING ZHONG • *Chongqing Research Center for Pharmaceutical Engineering, Chongqing Medical University, Chongqing, China*
SHUI ZOU • *Department of Pharmacy, National University of Singapore, Singapore, Singapore*

Part I

Bacterial Vectors

Chapter 1

Improving *E. coli* Bactofection by Expression of Bacteriophage ΦX174 Gene E

Dongwon Park, Andrew Hill, and Blaine A. Pfeifer

Abstract

Bactofection, a bacterial-mediated form of genetic transfer, is highlighted as an alternative mechanism for gene therapy. A key advantage of this system for immune-reactivity purposes stems from the nature of the bacterial host capable of initiating an immune response by attracting recognition and cellular uptake by antigen-presenting cells (APCs). The approach is also a suitable technique to deliver larger genetic constructs more efficiently as it can transfer plasmids of varying sizes into target mammalian cells. Given these advantages, bacterial vectors are being studied as potential carriers for the delivery of plasmid DNA into target cells to enable expression of heterologous proteins. The bacteria used for bactofection are generally nonpathogenic; however, concerns arise due to the use of a biological agent. To overcome such concerns, enhanced bacterial degradation has been engineered as an attenuation and safety feature for bactofection vectors. In particular, the ΦX174 lysis E (LyE) gene can be repurposed to both minimize bacterial survival within mammalian hosts while also improving overall gene delivery. More specifically, an engineered bacterial vector carrying the LyE gene showed improved gene delivery and safety profiles when tested with murine RAW264.7 macrophage APCs. This chapter outlines steps taken to engineer *E. coli* for LyE expression as a safer and more effective genetic antigen delivery bactofection vehicle in the context of vaccine utility.

Key words *E. coli*, Bactofection, ΦX174 Gene E, Lysis E (LyE), Gene delivery

1 Introduction

The efficiency of gene therapy relies on the ability to deliver genetic material in both a safe and efficient manner. Following the initial successes in transferring bacterial genes into mammalian cells, bactofection emerged as an alternative method of engineered delivery of DNA into human cells [1]. An underlying advantage of bactofection is the potential to leverage natural invasion and survival strategies developed by microbial entities in the native transfer of DNA and RNA.

An example is the cytosolic localization capabilities of *Listeria monocytogenes* through use of an endosome-perforating listeriolysin O (LLO) protein, which activates upon the development of the

Kumaran Narayanan (ed.), *Bio-Carrier Vectors: Methods and Protocols*, Methods in Molecular Biology, vol. 2211,
https://doi.org/10.1007/978-1-0716-0943-9_1,

phago-lysosome after bacterial uptake [2]. For the last two decades, clinical trials have advanced various types of bacterial vectors for the delivery of genes to certain cells for antiangiogenic, immunotherapy, and other therapeutic purpose [3–6]. Gene delivery using such nonviral bacterial vectors has been especially highlighted targeting antigen-presenting cells (APCs). In such a context, the bacterial vector possesses natural adjuvant features capable of triggering activation features of the APC through the production of nitric oxide to attract additional APCs to the site of infection and accelerate digestion of the bacteria [7]. To enhance endosomal escape after bacterial vector phagocytosis in APCs, LLO has been naturally leveraged or recombinantly introduced during bactofection [8].

While bactofection offers potential advantages in mammalian gene delivery, the approach still depends on a foreign bacterial cell, which can cause unwanted side effects and potential excesses in immune reactivity. Common bacterial agent attenuation approaches include chemical and biological-genetic cellular weakening [9]. This chapter will feature an alternative attenuation method used previously in vaccine-based bactofection strategies, namely introduction of the bacteriophage ϕX174 lethal lysis gene E (LyE) into *Escherichia coli* (*E. coli*) to promote safer and improved delivery of antigenic content to APCs [10]. The inclusion of LyE expression within the vector serves to both enhance bacterial degradation, thus, minimizing possible negative side effects caused by the bacterial cells, and improve release of DNA/RNA in APCs (in a similar function as LLO). Resulting bacterial vectors with the LyE gene have shown enhanced gene and protein release and inducible attenuation [11].

Figures 1 and 2 describe the mechanism of LyE activity affecting the cell wall and compares traditional bactofection with LyE integrated bactofection, respectively. The new LyE integrated strains exhibited improved gene delivery and reduced cytotoxicity profiles when tested with murine RAW 264.7 macrophage APCs. This chapter will describe the protocol used to engineer *E. coli* strains with the LyE gene and steps to assess membrane disruption, protein/DNA release, and cytotoxicity of the new bacterial vectors to serve as antigen delivery vehicles.

2 Materials

2.1 Molecular Biology Reagents and Equipment

1. Bacteriophage ΦX174, (ATCC).
2. Restriction enzymes: *Eco*RI, *Hin*dIII.
3. Primers.

   Forward: 5′-GGGAATTCGATGGTACGCTGGACTTTGTGG-3′

   Reverse: 5′-AGGAAGCTTTCACTCCTTCCGCACGTAATT-3′

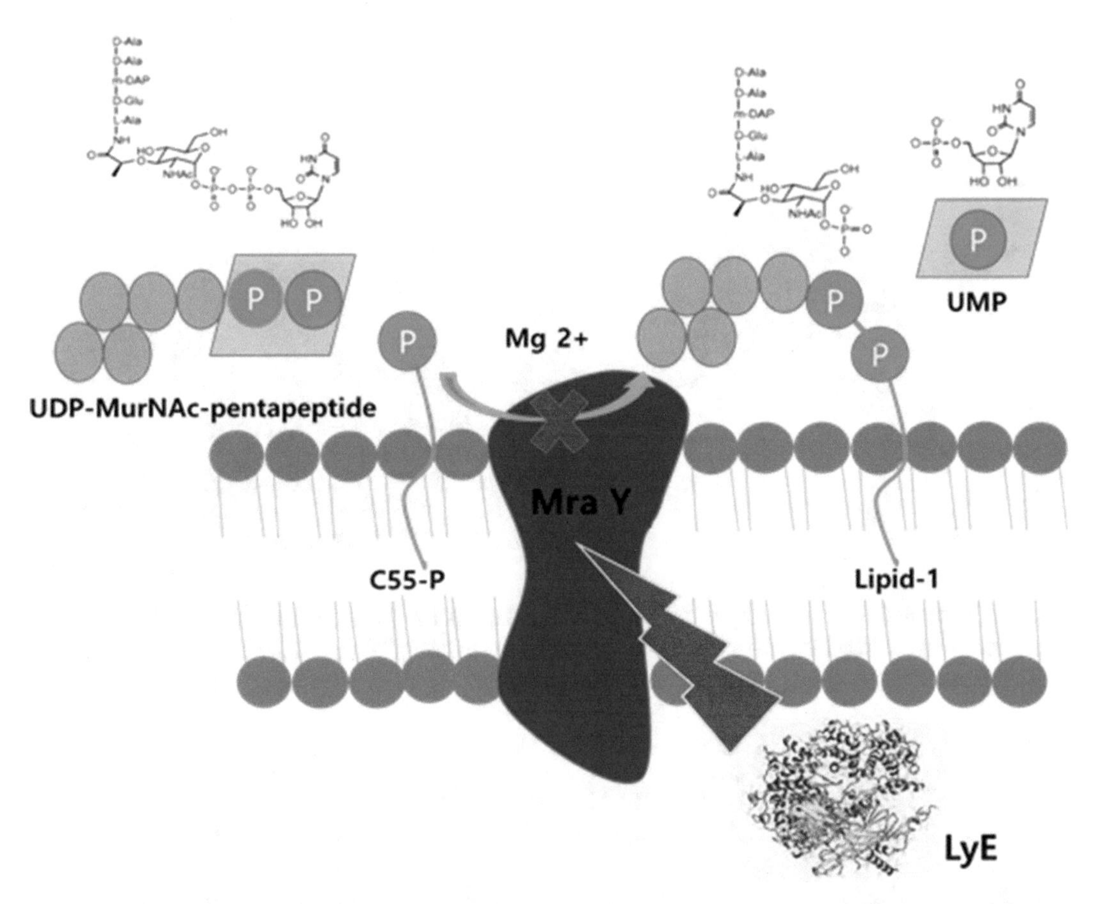

Fig. 1 Mechanism of LyE lysis. The Mra Y protein is responsible for the formation of the first lipid intermediate in cell wall biosynthesis. During lipid-1 formation, UDP-MurNac-pentapeptide attaches to the phosphate site of C55-P through Mra Y activity with the aid of magnesium. LyE inhibits the activity of MraY and generates immature lipids for cell wall biosynthesis resulting in the formation of transmembrane tunnels through the bacterial cellular envelope

4. Phusion DNA polymerase (New England Biolabs).
5. GeneJET Gel Extraction Kit (ThermoFisher Scientific).
6. pCMV-Luc: contains a luciferase reporter gene (Elim Biopharmaceuticals).
7. pACYC-Duet: cloning plasmid (EMD Millipore).
8. pET29-*hly*: plasmid containing the listeriolysin O gene (*hly*) under a T7 promoter [10].
9. MicroPulser electroporator (Bio-Rad Laboratories).
10. GENESYS 20 visible spectrophotometer (Thermo Fisher Scientific).
11. Branson 450D sonifier (Brandson Ultrasonic Corporation).
12. Synergy 4 multi-mode microplate reader (BioTek).

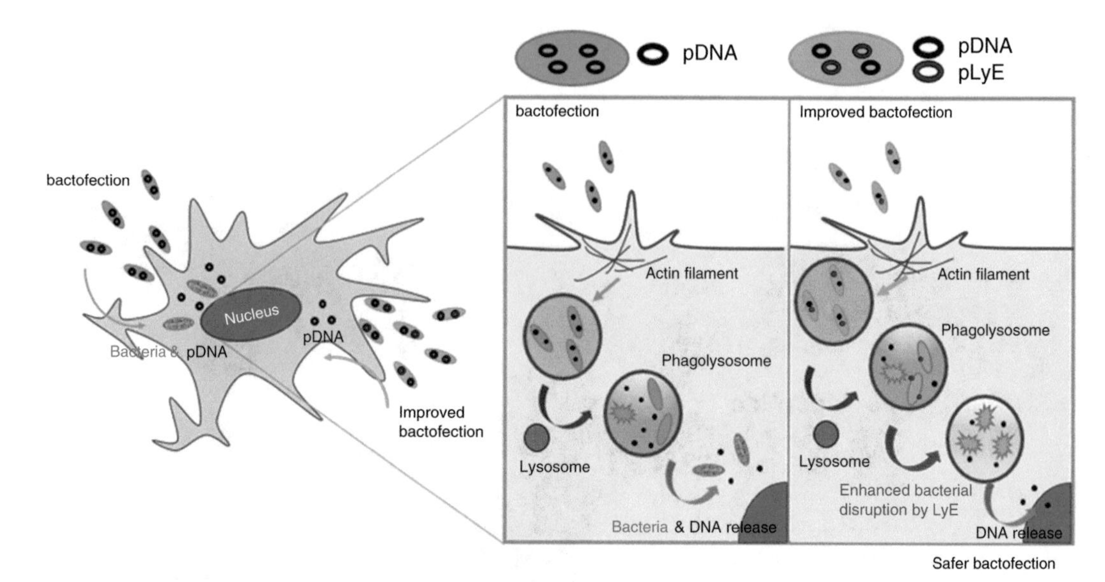

Fig. 2 LyE-mediated gene delivery in APC. The picture depicts two bactofection methods. The traditional bactofection vector (blue) may result in excess bacterial cells accumulating in the cell or in vivo, while the improved bactofection (green) enhances the bacterial disruption by expressing the LyE protein

13. Clear polystyrene 96-well tissue culture-treated microplate.
14. White tissue culture-treated 96-well plate.
15. AccuSpin Micro 17 Centrifuge (Fisher Scientific).

2.2 Bacterial Strains and Cell Lines

1. BL21(DE3): strain enabling recombinant protein expression using the T7 promoter (New England Biolabs).
2. YWT7-*hly* and YWTet-*hly*: strains contain the listeriolysin O gene (*hly*) integrated into the *E. coli* chromosome under either a T7 or Tet promoter, respectively [8]. Strains containing the *hly* gene further help improve bactofection gene delivery.
3. Murine macrophage line (RAW264.7) (ATCC).
4. Sheep red blood cells (Hemostat Laboratory).

2.3 Cell Growth Media and Reagents

1. Lysogeny broth (LB) media: 10 g/L tryptone, 5 g/L yeast extract, and 10 g/L NaCl. To generate LB agar for plate growth, LB media was supplemented with 20 g/L agar.
2. Medium components used for culture sterility and/or bacterial plasmid selection were included separately with the following concentrations: 50 mg/L kanamycin, 100 mg/L penicillin, 25 mg/L streptomycin, 50 mg/L gentamicin, and 100 mg/L ampicillin.
3. Phosphate buffered saline (PBS): 137 mM NaCl, 2.7 mM KCl, 10 mM Na_2HPO_4, 1.8 mM KH_2PO_4 at pH of 7.4 and 5.15.
4. Cysteine.

5. A stock of 200 mM of isopropyl β-D-1-thiogalactopyranoside (IPTG) for bacterial induction of gene expression.
6. A stock of 2 mg/mL L-arabinose for bacterial induction of gene expression.
7. Murine macrophage medium: 50 mL fetal bovine serum (heat inactivated), 5 mL 100 mM MEM sodium pyruvate, 5 mL 1 M HEPES buffer, 5 mL penicillin/streptomycin solution, and 1.25 g of D-(+)-glucose in 500 mL of phenol red-containing RPMI was mixed and sterile filtered through a 0.2 μm pore sized membrane.
8. MTT (3-(4,5-dimethylthiazol-2-yl)-2,5-diphenyl-2H-tetrazolium bromide).
9. Sodium dodecyl sulfate.
10. TrypLE Express enzyme solution (Thermo Fisher Scientific).
11. Bright Glo assay kit (Promega).
12. Dimethyl sulfoxide (DMSO).
13. Isotemp 637D incubator (Fisher Scientific; for bacterial growth on solid plates).
14. MaxQ 4000 Orbital incubator shaker (Fisher Scientific; for bacterial liquid culture).
15. Symphony Air Jacketed CO_2 incubator 5.3A (VWR; for mammalian cell culture).

3 Methods

3.1 Plasmid Preparation

1. Perform PCR for the LyE gene using the following primers: a forward primer of 5′- GGGAATTCGATGGTACGCTG GACTTTGTGG-3′ and a reverse primer of 5′-AGGAAGCTT TCACTCCTTCCGCACGTAATT-3′ yielding *Eco*RI and *Hin*dIII digestion sites at each end (*see* **Note 1**).
2. Subject the PCR-amplified LyE gene to electrophoresis using 100 V at constant amplitude with a 0.8% DNA agarose gel.
3. Purify PCR-amplified DNA using the GeneJet gel extraction kit following the manufacturer's instructions.
4. Digest purified DNA with *Eco*RI and *Hin*dIII for 1 h at 37 °C.
5. Following the restriction enzyme digestion, subject the DNA to electrophoresis as described above.
6. Purify digested DNA using a GeneJet gel extraction kit.
7. Ligate the digested DNA into *Eco*RI and *Hin*dIII digested pACYC-Duet to generate the pCYC-LyE plasmid (*see* **Note 2**).
8. Heat inactivate the ligation reaction by incubating the reaction mixture at 65 °C for 10 min.

9. Transform pCYC-LyE by electroporation into select *E. coli* bactofection strains. In this case, the plasmid was electroporated into strains BL21(DE3)/pET29-hly/pCMV-Luc, YWT7-hly/pCMV-Luc, and YWTet-hly/pCMV-Luc using a MicroPulser electroporator (*see* **Note 3**).

3.2 Evaluating Bacterial Growth Inhibition by LyE

1. Inoculate *E. coli* strains into 3 mL of lysogeny broth (LB) medium containing the appropriate antibiotics within an orbital shaker at 250 rpm and 37 °C and grow overnight for 16–20 h (*see* **Note 4).**
2. Transfer 1.5 mL of the seed culture into a 200 mL flask containing 50 mL of fresh LB media with the appropriate antibiotics and culture under the same conditions.
3. Quantify the cell density within the culture by measuring the optical density at 600 nm (OD_{600nm}) using a GENEYS 20 visible spectrophotometer. Continue taking measurements until an OD_{600nm} of 0.4–0.5 is reached (*see* **Note 5**).
4. Once an OD_{600nm} of 0.4–0.5 has been reached, add isopropyl β-D-1-thiogalactopyranoside (IPTG) at varying concentrations (0, 100, 500, 1000, 2000 μM) to introduce gene expression.
5. Following the introduction of IPTG, incubate the cells at different temperatures (22, 30, 37 °C). Over the course of 12 h, measure the OD_{600nm} every 30 min (Fig. 3) to collect data for growth curves for each strain containing different plasmids (*see* **Note 6**).

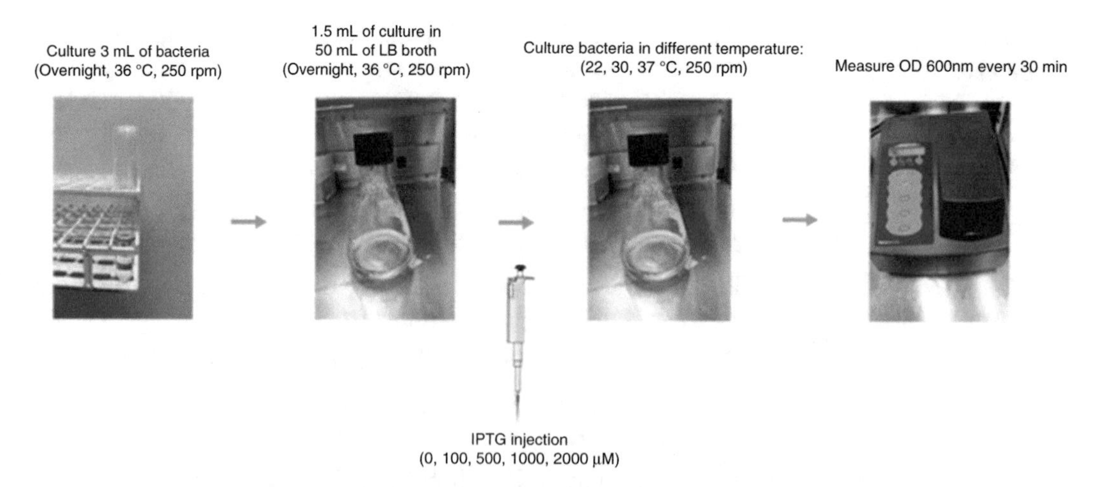

Fig. 3 Schematic of bacterial growth assay. Each bacterial vector is inoculated overnight followed by a medium transfer prior to gene expression induction with IPTG and continued incubation at varying temperatures during which cell growth is monitored

3.3 Evaluating Membrane Disruption by LyE

1. Conduct **steps 1–4** as outlined in Subheading 3.2 (*see* **Note 7**).
2. Transfer 200 μL of culture to a 1.5 mL Eppendorf tube every half hour postinduction over the course of 12 h.
3. Harvest cells via centrifugation at 4 °C and 13,000 rpm (using a microfuge) for 10 min.
4. Remove the supernatant and resuspend the cellular pellet in 1 mL phosphate buffered saline.
5. Lyse the cells via sonication using a Branson 450D sonifier (400 W, tapered microtip) at 20% capacity for 5 s. Include a sample that is not sonicated to enable determination of the percent of cells that are disrupted.
6. Following sonication, plate the cells on LB agar plates (diluting as needed to enable accurate single colony formation and counting) and incubate for 24 h at 37 °C.
7. Quantify cell disruption by counting colony forming units (CFUs) with comparison to *E. coli* without any introduced recombinant cellular disruption mechanism.

3.4 Quantifying Protein and DNA Release

1. Inoculate *E. coli* strains into 3 mL of lysogeny broth (LB) medium containing the appropriate antibiotics within an orbital shaker at 250 rpm and 37 °C and grow overnight for 16–20 h.
2. Transfer *E. coli* into 10 mL of fresh antibiotic-containing LB medium at 37 °C with 2.5% (v/v) inoculum from starter cultures.
3. Measure cellular density (via OD_{600nm} measurement) every hour until reaching an OD_{600nm} value of 0.4–0.5.
4. Induce gene expression through the addition of IPTG (100 μM) and incubate the culture at 22 °C for another hour. As a positive control, Polymyxin B (PLB), a bacterial cell wall disruption agent, should be added during IPTG induction to the final concentration of 0.5 mg/mL (*see* **Note 8**).
5. Standardize cultures for each strain to a 0.5 OD_{600nm} by measuring cellular density every 15 min (*see* **Note 9**).
6. After standardization, centrifuge the bacterial cultures at 4 °C at 13,000 rpm (using a microfuge) for 10 min.
7. Transfer the supernatant to a 15 mL centrifuge tube to measure absorbance: 260 nm for DNA and 280 nm for protein quantification with Genesys 20 plate reader (Fig. 4).
8. Save one milliliter of the supernatant and the bacterial pellet for the hemolysis assay. Both the supernatant and pellet can be stored at −20 °C until ready.

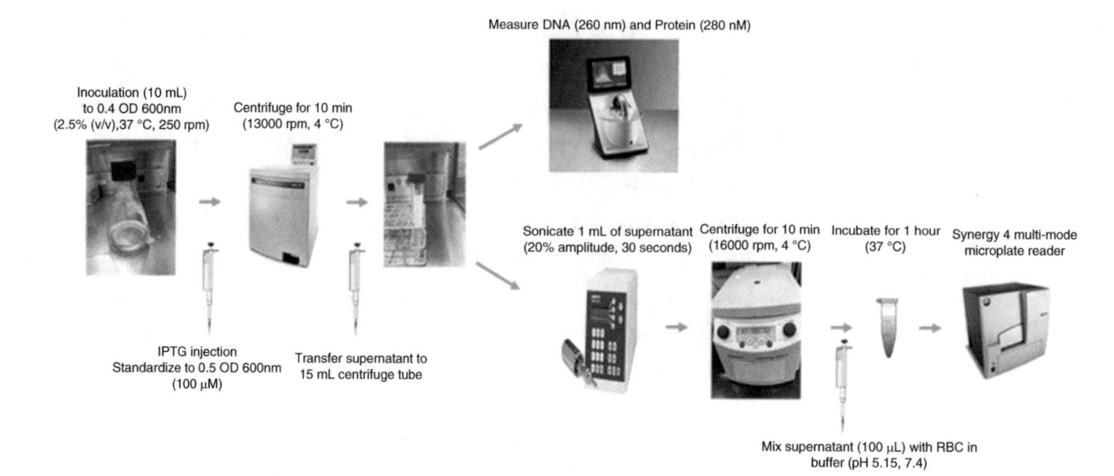

Fig. 4 Schematic of the macromolecule release and hemolysis assays. Bacterial vectors are prepared and standardized to an optical density value of 0.5 OD_{600nm} followed by assessment of passive release of macromolecular content (DNA and protein) and cellular capacity of internal components (such as listeriolysin O) capable of disrupting biological membranes (in this case, red blood cells [RBC])

3.5 Hemolysis Assay

The following assay is only conducted on bacterial vectors that contain listeriolysin O (LLO) and assesses the activity of recombinantly produced LLO via the ability to disrupt red blood cells (Fig. 4).

1. Sonicate the pellets saved during the DNA/protein release test described above after resuspending in 1 mL of PBS at an amplitude of 20% for 30 s with Branson 450D sonifier.
2. Following sonication, centrifuge the samples at 16,000 rpm using a microfuge for 10 min at 4 °C.
3. Following centrifugation, mix 100 μL of supernatant with 900 μL of 5% red blood cells in PBS buffer containing 6 mM cysteine at two separate pH values of 5.15 and 7.4. Include red blood cells that have not been mixed with bacterial lysate as a control for 0% hemolysis. Add sodium dodecyl sulfate (SDS) to the red blood cells to a concentration of 0.1% to achieve 100% hemolysis as a positive control.
4. Incubate the mixture at 37 °C for an hour.
5. Quantify LLO activity by measuring absorbance at 541 nm as compared with a standard (centrifuged 5% red blood cell in a corresponding PBS buffer) to calculate % red blood cell (RBC) lysis. The calculation to be used is as follows:

$$\%\text{Hemolysis} = (\text{Sample} - \text{negative control})/(\text{positive control} - \text{negative control}) * 100\%$$

3.6 Gene Delivery Assessment

1. Incubate RAW264.7 macrophage cells with murine microphage medium (Subheading 2.3) in T75 flasks and culture at 37 °C in a humidified incubator at a 5% CO_2 level until cells become confluent, passaging the medium and cells as needed.
2. Remove the medium via aspiration with a pipette and wash the cells with 1× volume of PBS buffer (pH 7.4) one time.
3. Detach the cells from the surface by incubating the cells with Express enzyme solution for 10 min.
4. To quench the trypsin reaction, add 10 mL of complete cell culture and mix by pipetting.
5. Measure the cell density using a hemocytometer (*see* **Note 10**).
6. Transfer the cells to a 15 mL conical tube and centrifuge at 500 × *g* for 5 min at 18 °C.
7. Remove the supernatant and then resuspend the pellet in antibiotic-free medium to reach 3×10^4 cells per 100 μL.
8. Transfer 100 μL of the cell solution into each well of two different types of 96-well plates (a clear, culture-treated format and a flat-bottomed, sterile, white polystyrene format) and incubate for 24 h at 30 °C.
9. Culture bacterial strains to be used for gene delivery as described for the DNA and protein release assay (10 mL media) and induce for an hour with 100 μM of IPTG at 22 °C.
10. Harvest bacterial cells via centrifugation at 4 °C and 13,000 rpm in a microcentrifuge for 10 min.
11. Wash the bacterial pellet washed using 10 mL of PBS.
12. Following the wash with PBS, standardize the cell density to an OD_{600nm} of 0.5 in PBS.
13. Dilute the bacterial cells in antibiotic-free RPMI-1600 in different ratios to create multiple bacteria-to-macrophage multiplicity of infection (MOI) ratios (1:1, 10:1, 100:1, 250:1, 500:1, 750:1, 1000:1, 1250:1, 1500:1, and 2000:1) (*see* **Note 11**).
14. Afterwards, remove the macrophage medium in both 96-well plates and replaced it with 50 μL of each respective bacterial-RPMI dilution (to generate the resulting MOI values) and incubate for an hour at 30 °C in a 5% CO_2 incubator. Also include a sample of macrophage which have no bacteria added as a negative control (*see* **Note 12**).
15. Add 50 μL of gentamicin-containing RPMI 1640 (100 mg/L) to each well and incubate for 24 h at 30 °C.
16. Quantify the luciferase expression using the Bright Glo assay kit (Promega) following the manufacturer's instructions.
17. Calculate gene delivery by dividing luciferase expression by protein content for each well/plate (Fig. 5).

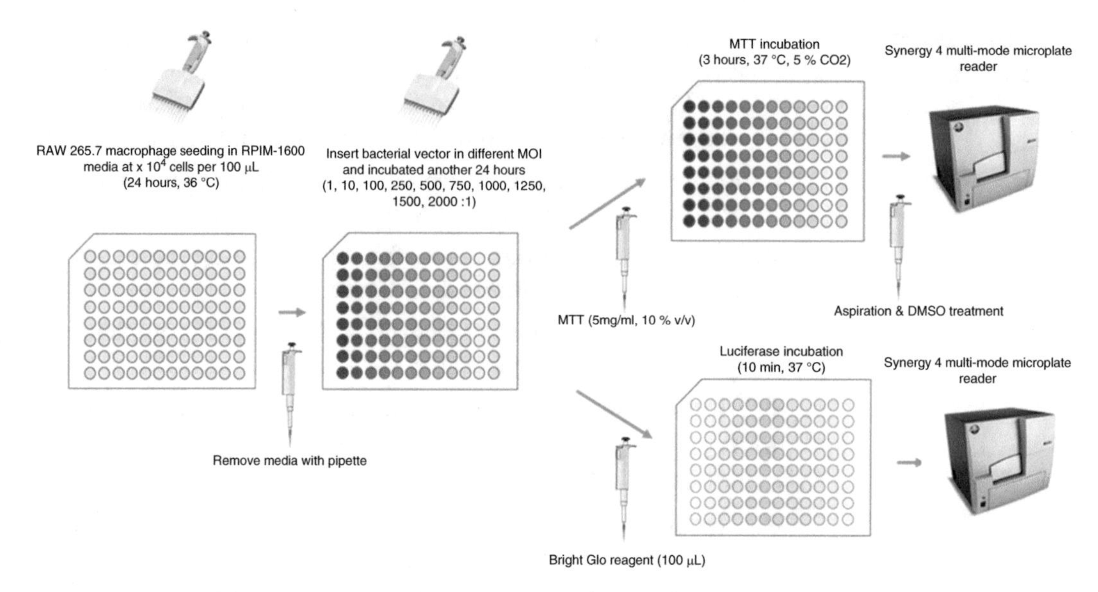

Fig. 5 Schematic of cytotoxicity and gene delivery assays. Macrophage cells are seeded prior to exposure to bacterial vectors at varying multiplicity of infection (MOI) ratios to then assess macrophage cellular health (and the impact such bacterial content as LyE has upon attenuation via an MTT assay) and gene delivery (assessed through luciferase gene delivery and macrophage cell expression)

3.7 Mammalian Cell Cytotoxicity of the E. coli Vector

1. Prepare murine macrophage cells (RAW264.7) as described in Subheading 3.6 following **steps 1–9**.
2. Prepare bacterial vectors combined with macrophage as described in Subheading 3.6 following **steps 10–14**.
3. Carefully remove the macrophage medium in both 96-well plates and replace it with 50 μL of each respective bacterial-RPMI dilution (to generate the resulting MOI values) and incubate at 30 °C in a 5% CO_2 incubator. Also include a sample of macrophage which have no bacteria added as a control.
4. After 24 h of incubation, add MTT (3-(4,5-dimethylthiazol-2-yl)-2,5-diphenyl-2H-tetrazolium bromide) solution (5 mg/mL) to each well at 10% (v/v) and incubate for 3 h at 37 °C and 5% CO_2. Include a negative control containing the cell culture medium alone. Also assemble varying solutions containing cell concentrations ranging from 10^3–10^6 cells per mL to establish a standard curve for the assay.
5. Aspirate the solution and resuspend the cells in 1× volume of dimethyl sulfoxide (DMSO) and incubate on a shaker for 1 h at 200 rpm agitation.
6. Measure the absorbance at 570 nm with a Synergy 4 multi-mode microplate reader (Fig. 5). Subtract the background using the absorbance of the aforementioned negative control. For a positive control, add 10% SDS to lyse the cells.

7. Utilize the standard curve described in **step 4** to determine the cell concentration and compare to the negative control to evaluate cytotoxicity (*see* **Note 13**).

4 Notes

1. PCR amplification is performed with Phusion DNA polymerase using the following conditions: 10 s at 98 °C, 30 s for primer annealing, 30 s per kb for extension repeated for 30 cycles.
2. Ligation reactions were carried out using 50 ng of vector DNA and an insert to vector ratio of 5:1. The room temperature at which the reactions were carried out was generally between 20 and 25 °C.
3. Since large plasmids may be used in bactofection studies, introduction of such plasmids to *E. coli* cells may require use of an alternative electro-transformation protocol instead of standard chemical transformation.
4. This growth inhibition experiment was performed to measure LyE expression impact upon *E. coli* growth. The expression level can be adjusted by IPTG concentration and temperature.
5. At the OD_{600nm} listed for this step (0.4–0.5), *E. coli* is experiencing the maximum rate of growth during log phase and is expected to be at peak cellular health.
6. The expected trend is that bacterial cellular viability will decrease as LyE levels increase, due to increased IPTG induction for example.
7. The membrane disruption assessment is to measure membrane disruption due to the LyE expression. As such, the strains were incubated initially under the same conditions as the previous subsection experiments.
8. Polymixin B acts as an effective antibacterial agent via cell membrane disruption and, thus, serves as a good comparative positive control in this case.
9. This step usually takes 1 h. However, a sufficient starting volume should be utilized to account for the amount of liquid removed while measuring optical density. We recommend using an additional 10 mL beyond the amount intended to ensure a sufficient quantity remains when an OD_{600nm} of 0.5 is achieved.
10. The cell density is measured to enable seeding into individual wells prior to transfection. Seeding levels can be adjusted as needed to maximize final signals but one must plan accordingly

during cell culture to ensure sufficient numbers of cells to conduct the transfection experiments across all the associated vectors and controls.

11. MOI indicates the ratio of bacteria to macrophage.
12. Due to the risk associated with disturbing the cells, we recommend removing the media via gentle aspiration. This can be achieved by attaching a sterile glass pipette to a vacuum line and carefully removing the media from a corner of the well or a similar location that is away from the majority of cells.
13. With a higher MOI, reduced macrophage viability is expected due to the greater bacterial load negatively influencing the coculture environment. Thus, through LyE expression, the experiment will reveal to what extent attenuation of the bacterial vectors affects this trend.

Acknowledgments

The authors recognize support from NIH awards AI088485 and AI117309 and a grant from the Technology Accelerator Fund from the State University of New York Research Foundation (all to BAP).

References

1. Courvalin P, Goussard S, Grillot-Courvalin C et al (1995) Gene transfer from bacteria to mammalian cells. C R Acad Sci III 318:1207–1212
2. Grillot-Courvalin C et al (1998) Functional gene transfer from intracellular bacteria to mammalian cells. Nat Biotechnol 16(9):862
3. Sizemore DR, Branstrom A, Sadoff J (1995) Attenuated Shigella as a DNA delivery vehicle for DNA-mediated immunization. Science 270:299–302
4. Grillot-Courvalin C, Goussard S, Courvalin P (2002) Wild-type intracellular bacteria deliver DNA into mammalian cells. Cell Microbiol 4:177–186
5. Al-Mariri A, Tibor A et al (2002) Yersinia enterocolitica as a vehicle for a naked DNA vaccine encoding *Brucella abortus* bacterioferritin or P39 antigen. Infect Immun 70:1915–1923
6. Gardlik R, Behuliak M, Palffy R, Celec P, Li C (2011) Gene therapy for cancer: bacteria-mediated anti-angiogenesis therapy. Gene Ther 18:425–431
7. Steinman RM (2001) Dendritic cells and the control of immunity: enhancing the efficiency of antigen presentation. Mt Sinai J Med 68:160–166
8. Birmingham CL, Canadien V, Kaniuk NA et al (2008) Listeriolysin O allows *Listeria monocytogenes* replication in macrophage vacuoles. Nature 451:350–354
9. Palffy R, Gardlik R, Hodosy J et al (2006) Bacteria in gene therapy: bactofection versus alternative gene therapy. Gene Ther 13:101–105
10. Chung TC, Jones CH, Gollakota A et al (2015) Improved *Escherichia coli* Bactofection and cytotoxicity by heterologous expression of bacteriophage PhiX174 lysis gene E. Mol Pharm 12:1691–1700
11. Parsa S, Wang Y, Fuller J (2008) A comparison between polymeric microsphere and bacterial vectors for macrophage P388D1 gene delivery. Pharm Res 25:1202–1208

Chapter 2

Improved DNA Delivery Efficiency of Bacterial Vectors by Co-Delivery with Exogenous Lipid and Antimicrobial Reagents

Andrew N. Osahor and Kumaran Narayanan

Abstract

Gene delivery using invasive bacteria as vectors is a robust method that is feasible for plasmid and artificial chromosome DNA construct delivery to human cells presenting β1 integrin receptors. This technique is relatively underutilized owing to the inefficiency of gene transfer to targeted cell populations. Bacterial vectors must successfully adhere to the cell membrane, internalize into the cytoplasm, undergo lysis, and deliver DNA to the nucleus. There are limited studies on the use of exogenous reagents to improve the efficiency of bacteria-mediated gene delivery to mammalian cells. In this chapter, we describe how cationic lipids, conventionally used for DNA and protein transfection, as well as antimicrobial compounds, can be used to synergistically enhance the adherence of invasive bacterial vectors to the cell membrane and improve their predisposition to internalize into the cytoplasm to deliver DNA. Using simple combinatorial methods, functional DNA transfer can be improved by up to four-fold of invaded cell populations. These methods are easy to perform and are likely to be applicable for other bacterial vectors including Listeria and Salmonella.

Key words Bactofection, Gene delivery, Invasion, Cationic lipid, Cell penetrating peptide, *E. coli*, Transfection efficiency

1 Introduction

Invasive bacteria capable of adherence to the cell membrane and internalization into the cytoplasm of mammalian cells can be used as vectors to deliver functional DNA. These vectors are typically capable of cloning transgenes during their replication cycles and protecting the DNA during the invasion process [1–3]. These vectors provide a potentially practical approach to circumvent issues associated with traditional vectors including immunogenicity, insertional mutagenesis packaging capacity, and production costs [4–6]. In addition to the delivery of oligonucleotides [1], bacterial vectors can be used for the in situ production of therapeutic proteins, vaccination, toxin binding, drug delivery, and enzyme pro-drug therapy [7–9].

Kumaran Narayanan (ed.), *Bio-Carrier Vectors: Methods and Protocols*, Methods in Molecular Biology, vol. 2211, https://doi.org/10.1007/978-1-0716-0943-9_2,

Unfortunately, the prospect of employing this class of vectors has been limited by low gene delivery efficiency [10, 11], while methods to improve the efficiency of the vector to render them clinically or physiologically viable remain poorly understood [12]. The vectors' ability to successfully adhere to the membrane of the target cell and internalize into the cytoplasm, which is directly correlated to its gene delivery efficiency remains fairly unsatisfactory, and is not yet a practical solution for molecule delivery in vivo [12].

There have been few successful attempts to improve viral and nonviral vector efficiency by co-delivery with exogenous reagents [13]. In these cases, peptides, oligonucleotides, and viral vectors were able to deliver reporter DNA with higher efficiency when they were complexed with cationic lipids. However, not many reagents have been investigated to aid the ability of bacterial vectors to deliver DNA efficiently. Antibiotics such as Polymyxin B have been shown to aid the bacterial invasion process by attenuating bacterial cells [10], and the transfection reagent Lipofectamine has been shown to have similar effects on the gene delivery efficiency of the vector [11].

Here we demonstrate that cationic lipids Lipofectamine and PULSin, which are commercially reagents employed for transfection and proteofection, respectively, as well as common antimicrobial reagents including amantadine, chloroquine, polymyxin B, and tetracycline [14, 15] can greatly enhance the efficiency of gene delivery using invasive *E. coli* as a vector (Fig. 1).

Gene delivery using bacterial vectors is typically performed in vitro by infection of the target/host cells, followed by incubation

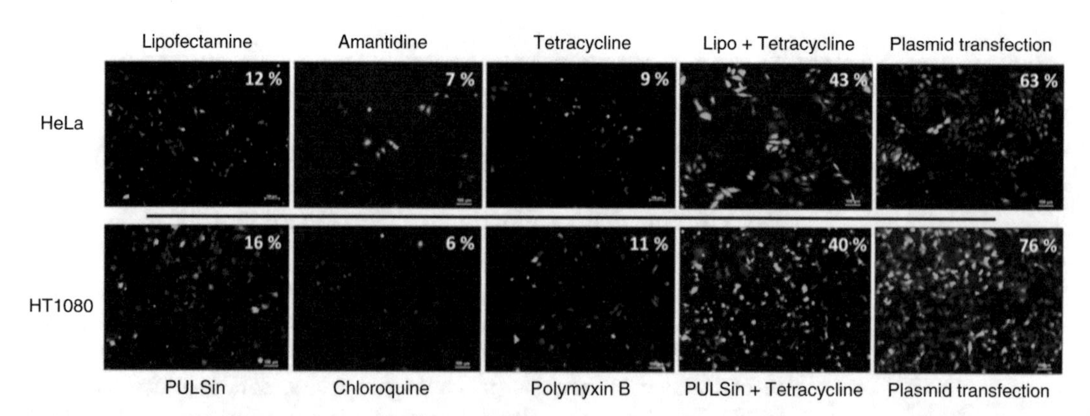

Fig. 1 Reporter gene delivery to HeLa and HT1080 cells using invasive *E. coli* complexed with lipids and antimicrobial reagents. Fluorescence images of HeLa and HT1080 cells captured at ×20 magnification 48 h after invasion using *E. coli* complexed with Lipofectamine, PULSin, amantadine, chloroquine, tetracycline, polymyxin B, and combinations of Lipofectamine/Tetracycline, and PULSin/chloroquine, compared to naked plasmid transfection using Lipofectamine 2000. White text indicates the percentage of reporter (GFP) expressing cells in the invaded population measured using FACS from a sample size of 40,000 cells

to allow successful adherence of the vector to the cell membrane and subsequent internalization into the cytoplasm [16]. Using this method, the invasive bacteria are simply mixed and incubated with lipids or antimicrobial reagents which are then co-delivered during invasion. Complexing these reagents with invasive *E. coli* prior to invasion of host cells provides vastly improved membrane adherence and internalization capabilities to the vectors with minimal toxicity to the targeted cell line (Figs. 2 and 3).

Using this method, we have recorded up to eight-fold increase in reporter expression in invaded cell populations when cationic lipids, and antimicrobial reagents are complexed with *E. coli* cells. Subsequently, we have observed that these reagents improve the vectors ability to deliver DNA to cells by greatly improving adherence to the cell membrane while simultaneously attenuating the vector, rendering them prone to easily release of their genetic cargo once inside the cell.

In this chapter, we describe methods to co-deliver invasive *E. coli* with selected lipids and antimicrobial reagents to enhance the gene delivery efficiency of this vector. This method is likely to be applicable for other bacterial vectors including Listeria and Salmonella, or for cell lines that are inherently difficult to target using bacterial vectors.

2 Materials

Prepare all solutions using ultrapure water and analytical grade reagents. Prepare and store all reagents at room temperature (unless indicated otherwise). Sterilize all reagents and culture vessels to avoid contamination. Diligently follow all waste disposal regulations when disposing waste materials.

2.1 Invasive E. coli Vector Preparation

1. Bacterial growth media: Brain Heart Infusion (BHI) media. Prepare according to manufacturer's instructions (*see* **Note 1**) and transfer to a 250 mL conical flask. Autoclave to sterilize. Store at room temperature.
2. Invasive *E. coli* vector: *E. coli* DH10B (pGB2Ωinv-hly, pEGFPN2) [17] (*see* **Note 2**).
3. Noninvasive control vector: *E. coli* DH10B (pEGFPN2) [17].
4. Cell lines: HEK 293 (ATCC #: CRL-1573), HeLa (ATCC #: CCL-2), HT1080 (ATCC #: CCL-121) (*see* **Note 3**).
5. Fluorescence-activated cell sorting: FACSCalibur flow cytometer (Beckton Dickinson) running CellQuest Pro V 6.0 software.

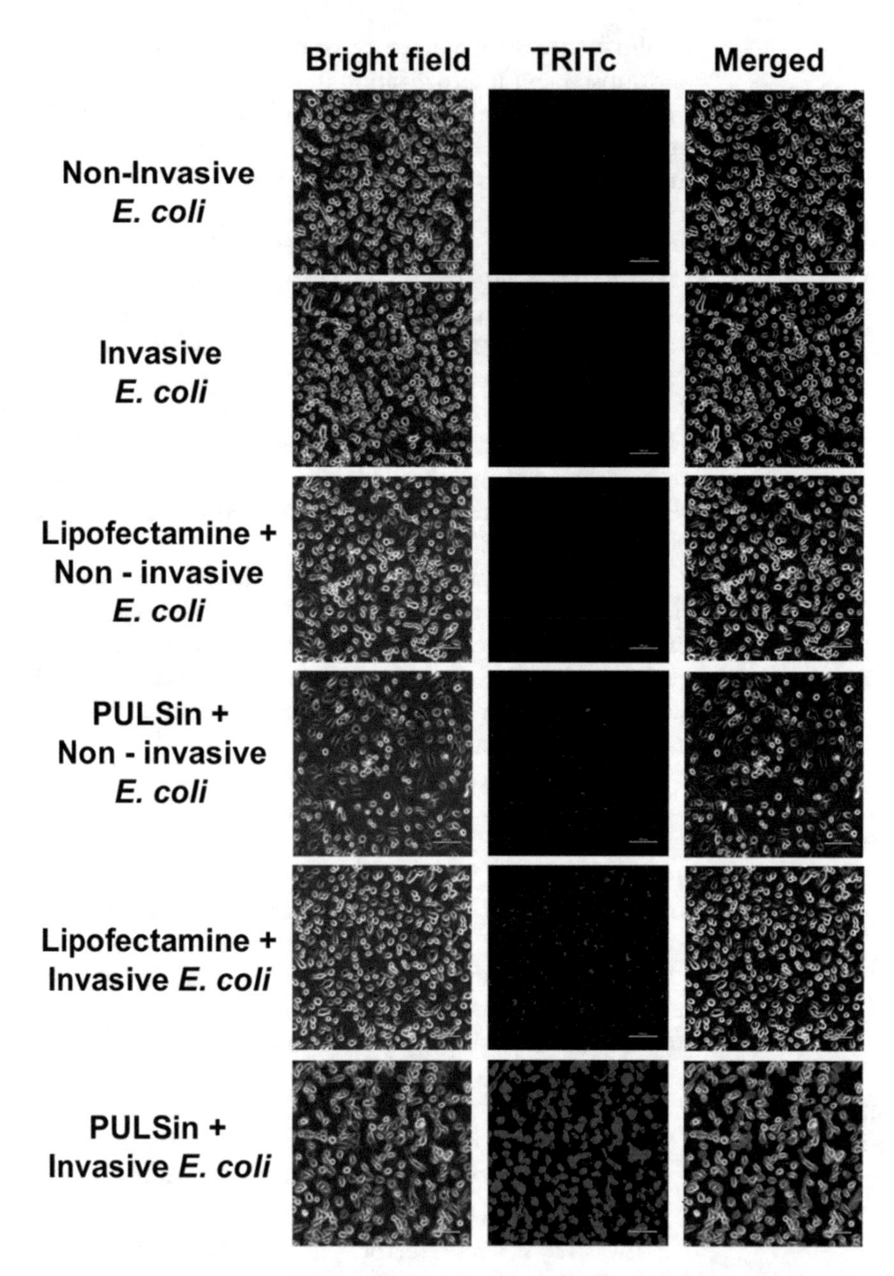

Fig. 2 Attachment of invasive *E. coli*–lipid complexes to the membranes of HT1080 cells. Cells invaded with *E. coli* complexed with Lipofectamine and PULSin observed using fluorescence microscopy at 20X magnification. Invasive *E. coli* (fluorescent red) expressing pUltraRFP-KM [18] uniformly adhere to cell membranes and are distributed throughout the invaded cell population in *E. coli*–lipid complexes, but not when invasive *E. coli* is used alone

2.2 E. coli Invasion Enhancing Reagents

1. Lipofectamine 2000: Store at 4 °C.
2. PULSin: Store at −20 °C.
3. 10 mM Amantadine hydrochloride: Dissolve 18.7 mg in 10 mL dH_2O (stock), working concentration of 50 μM (store at −20 °C).

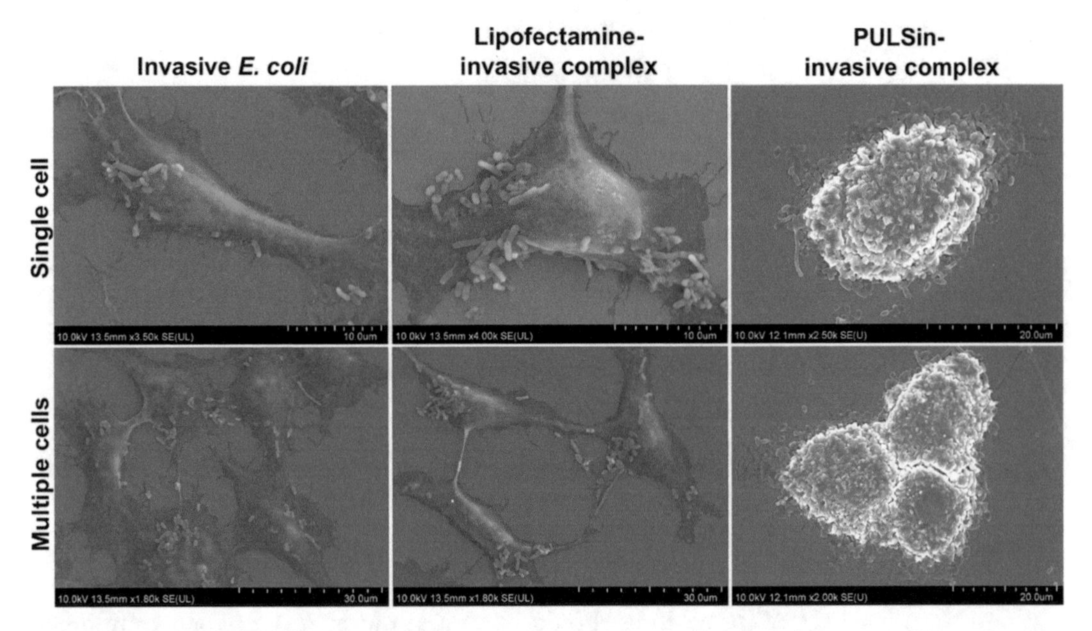

Fig. 3 Polar and diffuse attachment of *E. coli*–lipid complexes to HT1080 cell membranes. Electron microscope micrographs showing HT1080 fibroblasts (false colored pink) and invasive *E. coli* (false colored blue) in lipid directed arrangements onto individual cells and distributed in the same manner in multiple cells when Lipofectamine and PULSin are complexed with invasive *E. coli* to invade cells after 1 h of incubation, compared to the attachment of invasive *E. coli* alone without lipids. Images are qualitative representations of a three biological replicates for each treatment

4. 10 mM Chloroquine: Dissolve 51.6 mg in 10 mL dH_2O (stock), working concentration of 50 μM (store at −20 °C).
5. 12.5 mg/mL Tetracycline hydrochloride: Dissolve 12.5 mg in 10 mL in EtOH (stock), working concentration of 1 μg/mL (store at −20 °C). Keep away from light.
6. 10 mg/mL Polymyxin B sulfate: Dissolve 10 mg in dH_2O (stock), working concentration of 1 μg/mL (store at −20 °C).

2.3 Mammalian Cell Invasion

2.3.1 Cell Culture Media

1. Invasion media: RPMI or DMEM containing 10 mM L-glutamine. Store at 4 °C. Warm to 37 °C in a sterile water bath before use (*see* **Note 4**).
2. Complete culture media: RPMI or DMEM containing 10% Fetal Bovine Serum (FBS) and 10 mM L-glutamine. Store at 4 °C. Warm to 37 °C in a water bath before use (*see* **Note 5**).

2.3.2 Buffers and Reagents

1. Wash buffer: Phosphate Buffered Saline. Dilute stock from 10× to 1× using sterilized dH_2O. Store at 4 °C. Warm to 37 °C in a water bath before use.
2. 0.05% Trypsin ETDA.
3. 10 mg/mL Gentamicin. Store at 4 °C.

2.3.3 Cell Culture Materials

1. Six-well cell culture plates.
2. T-75 Cell culture flask.
3. 5 mL serological pipettes.
4. 10 mL serological pipettes.
5. Automatic pipettor.
6. Plate centrifuge.
7. Trypan blue.
8. Hemocytometer.
9. 15 mL Falcon centrifuge tubes.
10. 50 mL Falcon centrifuge tubes.
11. Water bath.

3 Methods

All procedures should be carried out at room temperature unless indicated otherwise.

3.1 Invasive *E. coli* Growth

1. Spread *E. coli* on freshly prepared BHI agar and incubate at 37 °C overnight until single colonies are obtained.
2. Inoculate 20 mL of sterile BHI broth with freshly grown single colonies of invasive *E. coli* DH10B (pGB2Ωinv-hly, pEGFPN2) and/or control noninvasive *E. coli* DH10B (pEGFPN2) (*see* **Note 6**).
3. Grow 10 mL of *E. coli* cultures overnight in 5 mL of BHI broth in a sterile 50 mL Falcon tube at 37 °C with shaking at 220 rpm overnight, or until late log/stationary phase is achieved approximately 18 h post-inoculation or until OD_{600} of 2.4 (*see* **Note 7**).
4. Harvest bacteria by centrifugation at 6000 × *g* for 7 min at room temperature. Discard the spent bacterial media by gently pouring out the supernatant (*see* **Note 8**).
5. Resuspend the bacterial pellet in 0.9 mL of invasion media and vortex thoroughly to ensure the bacterial pellet is fully dissociated and evenly mixed.

3.2 Mammalian Cell Growth

1. Transfer a vial of mammalian cells from cryopreservation and thaw rapidly in a 37 °C water bath (*see* **Note 9**).
2. Using a 5 mL serological pipette, gently remove cells from the cryopreservation vial into a T-75 cell culture flask containing complete cell culture media prewarmed to 37 °C in a 5% CO_2 incubator.

3. Incubate cells in a 37 °C incubator with 5% CO_2 with humidity overnight.
4. Wash cells with 1× PBS that has been prewarmed to 37 °C by gently pipetting onto the walls of the culture vessel and gently swirling to ensure proper washing. Transfer 15 mL of prewarmed complete culture media to the culture vessel. Grow cells until 70% confluent.
5. Aspirate spent media from the culture vessel using a serological pipette. Wash cells twice using 1× PBS by gently pipetting onto the walls of the culture vessel and gently swirling to ensure proper washing.
6. Remove 1× PBS using a serological pipette and transfer 4 mL of 0.05% trypsin to detach cells from the culture flask. Incubate cells in trypsin solution at 37 °C with 5% CO_2 and humidity for approximately 3–5 min or until cells are completely detached (*see* **Note 10**).
7. Gently add 5 mL of complete media to the culture vessel using a serological pipette to deactivate trypsin (*see* **Note 11**).
8. Collect detached cells into a 15 mL Falcon tube using a serological pipette. Centrifuge the cells at 225 × g for 5 min at room temperature.
9. Remove the supernatant using a serological pipette (*see* **Note 12**).
10. Resuspend the cell pellet in 1 mL of complete cell culture media. Pipette gently to ensure homogenous resuspension. Do not vortex.
11. Remove 10 μL of cells onto a hemocytometer using a micropipette and mix with 10 μL trypan blue dye to obtain a viable cell count.
12. Prepare cells for invasion by seeding HEK 293, HeLa, or HT1080 cells in 6-well plates at a density of 1×10^5 cells per well 24 h before invasion experiments (*see* **Note 13**).
13. On the day of invasion, remove growth media from the cell culture vessel using a sterile serological pipette and rinse cells twice with 1× PBS that has been prewarmed to 37 °C using a sterile water bath.
14. Remove 1× PBS from culture vessel using a serological pipette immediately before invasion (*see* **Note 14**).

3.3 Complexing E. coli with Invasion Enhancing Reagents

1. Dilute Lipofectamine 2000 and/or PULSin in invasion media that does not contain serum (*see* **Note 4**). Add 1 μL of either lipid to 30 μL of invasion media in a 1.5 mL centrifuge tube. Vortex thoroughly to ensure mixing. Incubate at room temperature for 10 min before complexation with invasive vector.

2. Transfer 10 μL bacteria that has been harvested during the late log growth phase and resuspended in invasion media into 30 μL of invasion media containing Lipofectamine 2000 or PULSin and vortex lightly or mix using a pipette. Incubate mixture at room temperature for 15 min before invading of host cells (*see* **Note 15**).
3. To include antimicrobial reagents to the invasive *E. coli*–lipid complex (*see* Fig. 1), dilute antibiotics including amantadine, chloroquine, tetracycline, and polymyxin B to their working concentration (*see* Subheading 2.2) in 950 μL of invasion media in a 1.5 mL centrifuge tube. Vortex thoroughly to mix. Incubate at room temperature for 10 min.
4. Transfer 40 μL of *E. coli*–lipid complexes (from Subheading 3.3, **step 1**) or 10 μL of harvested invasive *E. coli* (from Subheading 3.1, **step 5**) into the 1.5 mL centrifuge tube containing diluted antibiotics immediately prior to invasion of host cells. Mix thoroughly by pipetting.

3.4 E. coli Invasion and Co-delivery with transfection Reagents

1. Top up the mixture containing invasive *E. coli*–lipid complexes or *E. coli*–lipid–antibiotic solution to 1 mL using invasion media and mix by pipetting (*see* **Note 16**).
2. Wash cells twice by gently pipetting 3 mL of 1× PBS into 6-well plates and gently swirling to mix. Remove PBS using a serological pipette (*see* **Note 17**).
3. Transfer 1 mL of *E. coli*–lipid complexes from the 1.5 mL microcentrifuge tube onto the monolayer of cells in each well. Swirl plates gently to mix.
4. Centrifuge 6-well plates in a plate centrifuge at 225 × *g* for 10 min. Immediately transfer plates to a 5% CO_2 incubator set at 37 °C with humidity and incubate for 1 h (*see* **Note 18**).
5. Aspirate media from wells using a serological pipette. Gently rinse each well with 3 mL of 1× PBS (*see* **Note 19**).
6. Add 1 mL of complete cell culture media containing 80 μg/mL of gentamicin to each well. Incubate for 1 h at 37 °C in a CO_2 incubator with humidity.
7. After incubation, repeat **step 6** three times to remove gentamicin.
8. Add 1 mL of complete cell culture media containing 20 μg/mL of gentamicin. Incubate at 37 °C in a CO_2 incubator with humidity for 24–48 h to analyze gene expression (*see* **Note 20**).

3.5 Detecting Gene Expression

1. Remove complete culture media containing 20 μg/mL gentamicin using a serological pipette. Rinse cells 2–3 times using 1× PBS that has been prewarmed to 37 °C using a water bath.

2. Detach cells from plates by adding 0.5 mL of 0.05% trypsin per well and incubate in a 5% CO_2 incubator at 37 °C for 3–5 min (*see* **Note 10**).
3. Add 0.5 mL of complete cell culture media to each well to deactivate trypsin and transfer the contents of each well into a 15 mL Falcon tube.
4. Centrifuge the 15 mL tubes at 225 × *g* for 10 min at room temperature.
5. Gently aspirate the supernatant using a serological pipette ensuring that the cell pellet is not disturbed.
6. Resuspend the cell pellet by transferring 1 mL of 1× PBS into the 15 mL Falcon tube using a micropipette. Repeatedly pipette gently to ensure thorough resuspension (*see* **Note 21**).
7. Store the resuspended cells on ice away from light until they are analyzed using a flow cytometer (*see* **Note 22**).
8. Analyze cells for fluorescence resulting from successful reporter gene expression (*see* **Note 23**).

4 Notes

1. Reconstituted BHI growth media should always be sterilized. Media can be prepared and stored for future experiments but is recommended to be made fresh.
2. Plasmid pEGFPN2 can be substituted with any preferred eukaryotic expression vector or reporter construct.
3. Other cell lines that express β1 integrins on the cell membrane may be permissive for use with this invasive *E. coli* vector.
4. The base growth media used for each experiment is cell line specific. Most immortalized cell lines are maintained in RPMI or DMEM. It is important that media used for invasion does not contain serum as this leads to clumping of the bacterial vectors, which drastically reduces their invasion capabilities.
5. Complete cell culture media used to promote recovery of cells following the invasion process should contain all necessary components including 2 mM L-glutamine and 10% FBS. It is helpful to increase the serum concentration to 12% following the invasion process to improve recovery rates and limit cell death.
6. A sterile pipette tip or inoculation loop should be used to transfer colonies from the plate of BHI agar into liquid BHI broth. Aeration can be provided by leaving the caps of the 50 mL centrifuge tube slightly open. Culture tubes should be slightly slanted during incubation.

7. It is important to use fresh cultures of bacteria at the late log phase of growth for invasion. Prior analysis of strain-specific growth profiles may be required to determine appropriate bacterial counts for the desired multiplicity of infection (MOI) during invasion. When bacteria are harvested, they should be kept at room temperature throughout harvesting and during complexation with any exogenous reagents. Placing bacterial cultures on ice or at 4 °C will significantly inhibit the vectors ability to bind to the cell membrane during invasion.
8. Observe the bacterial pellet carefully to ensure that it does not dissociate while the supernatant is removed. The 50 mL tube can be inverted on tissue briefly for leftover BHI media to be absorbed.
9. Vials of cryopreserved cells should be kept on ice when they are transferred between liquid nitrogen storage and a water bath. Thawed cells that have been transferred into complete culture media should be pelleted and resuspended in fresh complete culture media to remove traces of cyropreservants like glycerol or DMSO before incubation overnight. Omitting this step may significantly reduce the viability of the thawed cells.
10. Trypsinization rate will vary slightly depending on the cell line that is used. The flask can be tapped lightly to detach cells. A 5 mL serological pipette can be used to gently wash off cells using the trypsin contained in the flask. Cells should be observed under the microscope to ensure that proper detachment from the cell culture flask is achieved.
11. Exposing cells to trypsin for too long may negatively impact the receptors present on the cell membrane. Trypsin should be inhibited using complete media containing serum immediately cells are sufficiently detached.
12. Ensure that the cell pellet is not disturbed while discarding the spent media supernatant. If the pellet is disturbed, additional centrifugation for 5 min at 225 × *g* can be performed.
13. The density of cells seeded for invasions will vary according to cell type. Cells should ideally be seeded at approximately 60% confluency. Slow growing cells can be seeded at higher densities provided crowding does not occur. When cells are incubated following the invasion process, non-exponential growth of the cells resulting from high confluency will limit the maximum achievable gene delivery efficiency. Cells should not be seeded for invasion a day after thawing from cryopreservation and should be allowed sufficient time to multiply until approximately 80% confluency.
14. The duration that cells are exposed to PBS or invasion media which do not contain serum should be minimized. Cells will begin to round or shrink the longer they are left in PBS or

serum free media. In addition, all traces of PBS should be removed to avoid diluting the subsequently used invasion media.

15. We find that extensive vortexing bacterial cultures after they have been complexed with cationic reagents results in significant lysis of bacteria. This can result in false-positive gene transfer resulting from free plasmids complexing with lipids gaining access to cells.
16. In experiments that test multiple variables, 1.5 mL centrifuge tubes containing mixtures of bacteria and reagents in invasion media can become numerous and tedious to sort through. We have found that the effects of these external reagents are not altered if the reaction is scaled up into a 15 mL centrifuge tube up to a maximum volume of 7 mL followed by aliquoting accordingly into 6-well plates containing cells.
17. Wash steps should be executed carefully as some cell types are prone to detachment. PBS should be gently pipetted on the sides of the well and swirled carefully. Shrinkage/rounding of cells occurs if PBS is not maintained at 37 °C. Wash time should be kept to a minimum and cells should not be exposed to PBS for too long. We have found it is helpful in some cases to condition cells to multiple washes with PBS during routine maintenance/passaging to minimize detachment when washing during invasion.
18. Cells can be quickly viewed under a microscope after centrifugation to observe the attachment characteristics of bacteria to the cell membrane.
19. Cells can be washed more than three times until all unattached bacteria are removed. Invaded cell populations can be quickly viewed under a microscope to determine is washing post-invasion is sufficient.
20. Gene expression can be observed approximately 24 h after invasion. The maximum amount of gene expression however is measurable between 48 and 72 h.
21. Cells should be thoroughly resuspended to prevent cell clumps from blocking the fluidic systems of the flow cytometer.
22. Cells can be left on ice for up to 3 h away from direct light before fluorescence signals begin to decrease. If cells are not analyzed immediately, they should be resuspended again immediately before they are analyzed using the flow cytometer.
23. A group of control cells that were not invaded with *E. coli* can be used to establish a baseline for fluorescence measurements in treated samples. We have found that fluorescence readings are more accurate when control cells are subjected to the same

treatments including treatment with invasive media, PBS washes, and gentamicin treatment, without invasion using *E. coli*.

Acknowledgments

This work was supported by a MOSTI eScience grant 02-02-10-SF0252 from the Ministry of Science, Technology and Innovation, Malaysia, to K.N. A.N.O. was supported by a Higher Degree by Research Scholarships from Monash University Malaysia.

References

1. Celec P, Gardlik R (2017) Gene therapy using bacterial vectors. Front Biosci (Landmark Ed) 22:81–95
2. Guzman-Herrador DL, Steiner S, Alperi A, Gonzalez-Prieto C, Roy CR, Llosa M (2017) DNA delivery and genomic integration into mammalian target cells through type IV a and B secretion systems of human pathogens. Front Microbiol 8:1503. https://doi.org/10.3389/fmicb.2017.01503
3. Cronin M, Stanton RM, Francis KP, Tangney M (2012) Bacterial vectors for imaging and cancer gene therapy: a review. Cancer Gene Ther 19(11):731–740. https://doi.org/10.1038/cgt.2012.59
4. Dalby B, Cates S, Harris A, Ohki EC, Tilkins ML, Price PJ, Ciccarone VC (2004) Advanced transfection with Lipofectamine 2000 reagent: primary neurons, siRNA, and high-throughput applications. Methods 33(2):95–103. https://doi.org/10.1016/j.ymeth.2003.11.023
5. Huh SH, Do HJ, Lim HY, Kim DK, Choi SJ, Song H, Kim NH, Park JK, Chang WK, Chung HM, Kim JH (2007) Optimization of 25 kDa linear polyethylenimine for efficient gene delivery. Biologicals 35(3):165–171. https://doi.org/10.1016/j.biologicals.2006.08.004
6. Palffy R, Gardlik R, Hodosy J, Behuliak M, Resko P, Radvansky J, Celec P (2006) Bacteria in gene therapy: bactofection versus alternative gene therapy. Gene Ther 13(2):101–105. https://doi.org/10.1038/sj.gt.3302635
7. Lehouritis P, Springer C, Tangney M (2013) Bacterial-directed enzyme prodrug therapy. J Control Release 170(1):120–131. https://doi.org/10.1016/j.jconrel.2013.05.005
8. da Silva AJ, Zangirolami TC, Novo-Mansur MT, Giordano Rde C, Martins EA (2014) Live bacterial vaccine vectors: an overview. Braz J Microbiol 45(4):1117–1129
9. Souders NC, Verch T, Paterson Y (2006) In vivo bactofection: listeria can function as a DNA-cancer vaccine. DNA Cell Biol 25 (3):142–151. https://doi.org/10.1089/dna.2006.25.142
10. Jones CH, Rane S, Patt E, Ravikrishnan A, Chen CK, Cheng C, Pfeifer BA (2013) Polymyxin B treatment improves bactofection efficacy and reduces cytotoxicity. Mol Pharm 10 (11):4301–4308. https://doi.org/10.1021/mp4003927
11. Narayanan K, Lee CW, Radu A, Sim EU (2013) *Escherichia coli* bactofection using lipofectamine. Anal Biochem 439(2):142–144. https://doi.org/10.1016/j.ab.2013.04.010
12. Hill AB, Chen M, Chen CK, Pfeifer BA, Jones CH (2016) Overcoming gene-delivery hurdles: physiological considerations for nonviral vectors. Trends Biotechnol 34(2):91–105. https://doi.org/10.1016/j.tibtech.2015.11.004
13. Hart SL, Arancibia-Carcamo CV, Wolfert MA, Mailhos C, O'Reilly NJ, Ali RR, Coutelle C, George AJ, Harbottle RP, Knight AM, Larkin DF, Levinsky RJ, Seymour LW, Thrasher AJ, Kinnon C (1998) Lipid-mediated enhancement of transfection by a nonviral integrin-targeting vector. Hum Gene Ther 9 (4):575–585. https://doi.org/10.1089/hum.1998.9.4-575
14. Debaisieux S, Rayne F, Yezid H, Beaumelle B (2012) The ins and outs of HIV-1 tat. Traffic 13(3):355–363. https://doi.org/10.1111/j.1600-0854.2011.01286.x
15. Pieper GM, Olds CL, Bub JD, Lindholm PF (2002) Transfection of human endothelial cells with HIV-1 tat gene activates NF-kappa B and enhances monocyte adhesion. Am J Physiol Heart Circ Physiol 283(6):H2315–H2321. https://doi.org/10.1152/ajpheart.00469.2002

16. Grillot-Courvalin C, Goussard S, Huetz F, Ojcius DM, Courvalin P (1998) Functional gene transfer from intracellular bacteria to mammalian cells. Nat Biotechnol 16 (9):862–866. https://doi.org/10.1038/nbt0998-862
17. Narayanan K, Warburton PE (2003) DNA modification and functional delivery into human cells using *Escherichia coli* DH10B. Nucleic Acids Res 31(9):e51
18. Mavridou DA, Gonzalez D, Clements A, Foster KR (2016) The pUltra plasmid series: a robust and flexible tool for fluorescent labeling of Enterobacteria. Plasmid 87–88:65–71. https://doi.org/10.1016/j.plasmid.2016.09.005

Chapter 3

Visualization of Bacteria-Mediated Gene Delivery Using High-Resolution Electron and Confocal Microscopy

Andrew N. Osahor, Allan Wee Ren Ng, and Kumaran Narayanan

Abstract

Visual analysis of the gene delivery process when using invasive bacteria as a vector has been conventionally performed using standard light and fluorescence microscopy. These microscopes can provide basic information on the invasiveness of the bacterial vector including the ability of the vector to successfully adhere to the cell membrane. Standard microscopy techniques however fall short when finer details including membrane attachment as well as internalization into the cytoplasm are desired. High-resolution visual analysis of bacteria-mediated gene delivery can allow accurate measurement of the adherence and internalization capabilities of engineered vectors. Here, we describe the use of scanning electron microscopy (SEM) to directly quantify vectors when they are external to the cell wall, and confocal microscopy to evaluate the vectors when they have internalized into the cytoplasm. By performing the invasion procedure on microscope coverslips, cells can be easily prepared for analysis using electron or confocal microscopes. Imaging the invasion complexes in high resolution can provide important insights into the behavior of bacterial vectors including *E. coli, Listeria, and Salmonella* when invading their target cells to deliver DNA and other molecules.

Key words Cell membrane, Invasion, Electron microscopy, Confocal microscopy, Cell line, Fluorescence

1 Introduction

Invasion of mammalian cells using invasive *Escherichia coli* (*E. coli*) as a vector has continually been developed as a means to safely deliver genes and therapeutic molecules [1]. Historically, successful gene transfer to invaded cells is usually quantified via eukaryotic expression of fluorescent proteins like Green Fluorescent Protein (GFP) using standard fluorescence microscopy or Fluorescence Activated Cell Sorting (FACS). These tools have proven sufficient to determine transfection efficiencies during invasion experiments [2–4], but do little to elucidate key interactions between the bacterial vectors and their target cells including vector–membrane interactions and the ability of the vector to internalize into the cytoplasm.

Kumaran Narayanan (ed.), *Bio-Carrier Vectors: Methods and Protocols*, Methods in Molecular Biology, vol. 2211, https://doi.org/10.1007/978-1-0716-0943-9_3,

For analysis of the internalization of bacteria into the cytoplasm, invaded cell populations are typically subjected to the gentamicin protection assay [5, 6]. While this method is widely adapted for the analysis of cellular internalization, it may be flawed considering the potential for leakage of gentamicin into eukaryotic cells. It is also possible that the eradication of bacterial vectors external to the cell membrane by gentamicin may not be thorough. Minor variations to these bacterial internalization and transgene expression assays have been successfully utilized [1, 7]; however, there is no consensus regarding the accuracy in which invasion outcomes are reported, especially including successful membrane attachment and internalization into the cytoplasm.

High-resolution imaging can be applied as an alternative to accurately evaluate these interactions of bacterial vectors with targeted cells. SEM is a far superior alternative to resolve interactions of bacterial vectors external to the cell membrane and supplies enough resolution to virtually enable manual quantification of individual successful vector attachment events. This tool is particularly useful to visualize bacteria–host interactions when invasive bacteria are complexed with exogenous reagents [8, 9] (Fig. 1). Correspondingly, confocal microscopy may be a preferable solution to analyze internalization into the cytoplasm compared to the

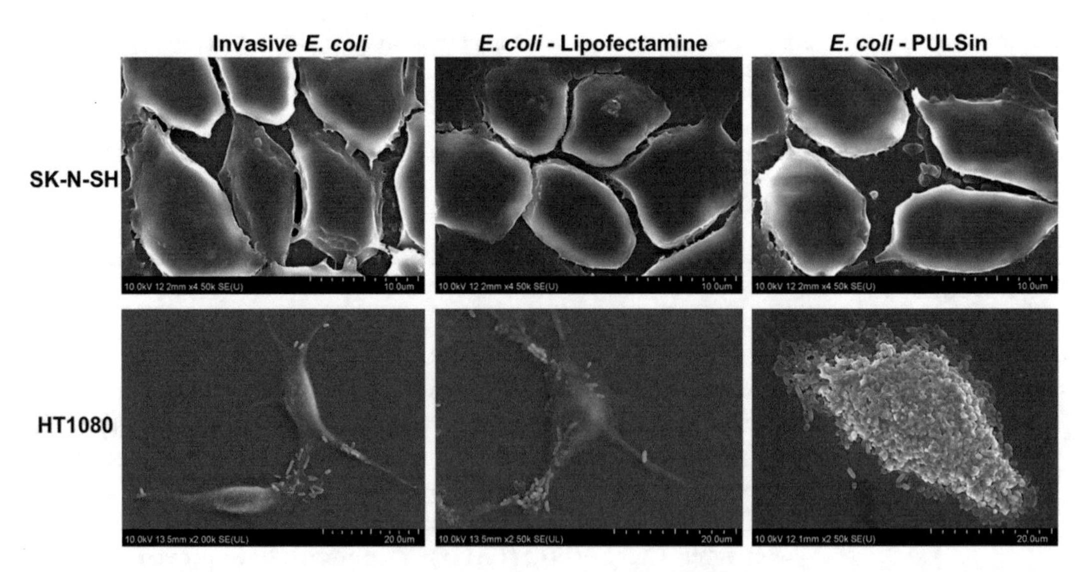

Fig. 1 Micrographs of human cells invaded with engineered *E. coli*. Cells were grown and invaded on plastic coverslips and prepared for scanning electron microscopy by fixation and dehydration. Micrographs show β1 Integrin-specific vector adherence of *E. coli*–lipid complexes to human cells. Neuroblastoma (SK-N-SH) that do not express membrane β1 integrins were invaded with invasive *E. coli*–lipid complexes. Compared to fibrosarcoma (HT1080) cells that express abundant β1 integrin. FESEM micrographs show SK-N-SH at ×4500 magnification was not receptive to membrane adherence, even when cationic lipids were complexed with invasive *E. coli*. HT1080 cells at ×2000–2500 magnification allowed higher vector attachment of *E. coli*–lipid complexes

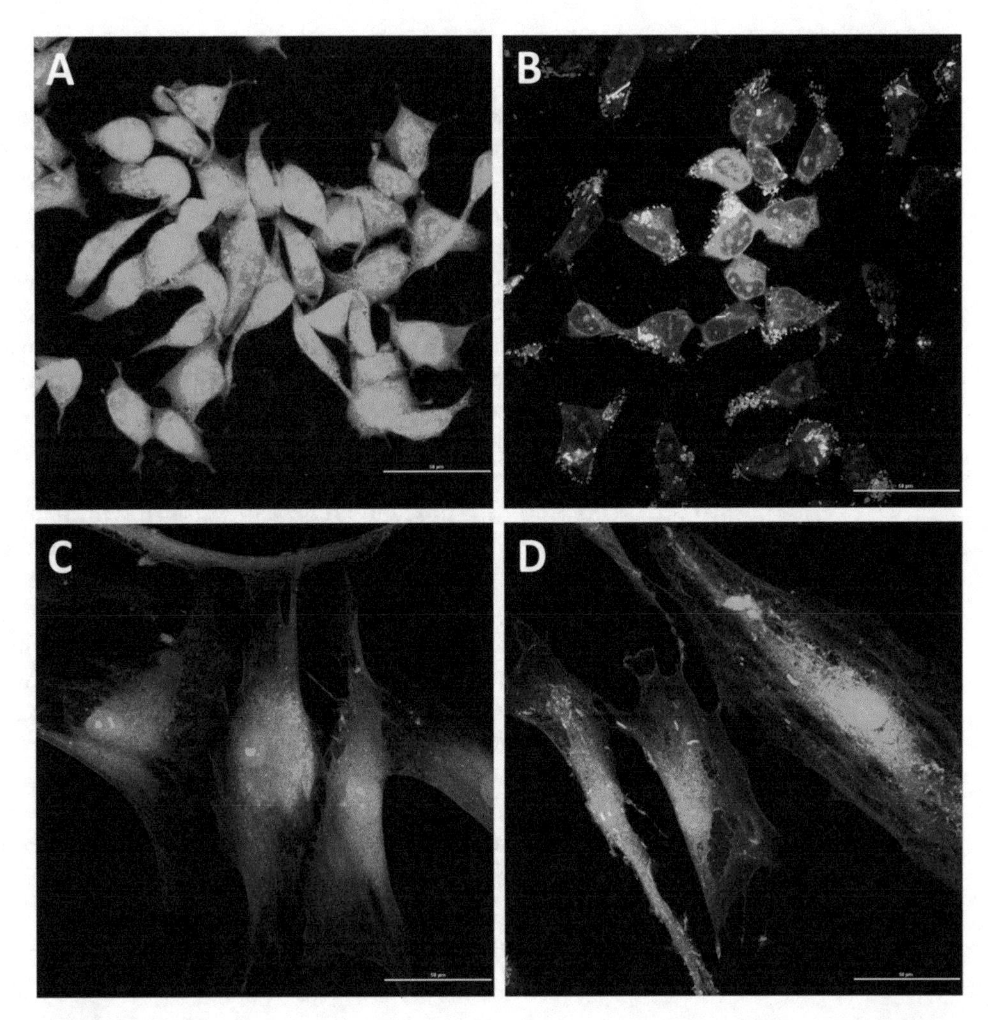

Fig. 2 Internalization of invasive *E. coli* into the cytoplasm. Visual appearance of microbes on host cells grown and invaded on glass coverslips captured using confocal microscopy. Cells were fluorescently stained (green) using acridine orange and invasive *E. coli* constitutively express RFP (Red). The cancer fibroblasts in (**a**) express the targeted receptor in high amounts and (**b**) Bind and internalize more bacteria as a result (yellow). By comparison, the noncancer fibroblasts in (**c**), express less of this receptor and as a result (**d**) have less microbial interaction with the cell membrane and no observable internalization into the cytoplasm

standard gentamicin protection assay [10]. Visualizing internalized bacteria and its progress within cells using confocal microscopy supplies a much clearer picture of vectors' abilities to successfully internalize into the cytoplasm (Fig. 2), but is not used routinely when studying molecule delivery using bacteria.

Here, we demonstrate the use of high-resolution microscopy including Scanning Electron Microscopy (SEM) and confocal microscopy to visualize the attachment and internalization characteristics of invasive *E. coli* during the invasion of human cells. These techniques can be easily and cost-effectively adapted for analysis of the invasion process [9], providing clear details of successful vector attachment events on the target cell membrane, and well-defined indications of the internalization of the vector into the cytoplasm.

Culturing cells on microscope coverslips inserted into typical culture vessels allows invaded cell populations to be prepared for advanced microscopy with minimal effort. Using SEM, the interactions of invasive vectors with the cell membrane can be studied with high enough resolution to allow the quantification of individual successful bacterial attachment events to the cell. This procedure can be performed without the use of hazardous and expensive fixatives such as osmium tetroxide or hexamethyldisilane, as well as critical point or freeze drying [9].

Invaded populations on culture slips can also be subjected to confocal microscopy to clearly determine vector internalization into the cytoplasm. This process can be achieved using inexpensive fluorescent stains meant for DNA staining such as acridine orange [11] and propidium iodide [12] to demarcate the cell membrane and nucleus, allowing distinction between vectors constitutively expressing fluorescent proteins that are external to the cell membrane, and vectors that have been internalized into the cytoplasm.

These techniques take advantage of easy preparatory steps and affordable materials to appropriately visualize important aspects of the invasion process when bacteria are used as vectors for molecule delivery. These methods may be applicable to precisely evaluate the effectiveness of the membrane adherence and internalization characteristics of other bacterial vectors.

2 Materials

Prepare all solutions using ultrapure water and analytical grade reagents. Prepare and store all reagents at room temperature (unless indicated otherwise). Sterilize all reagents and culture vessels to avoid contamination. Diligently follow all waste disposal regulations when disposing waste materials.

2.1 Buffers and Reagents

1. Wash buffer: Phosphate Buffered Saline (PBS). Dilute stock from 10× to 1× using sterilized dH_2O. Store at 4 °C. Warm to 37 °C before use during bacterial invasion procedures.
2. Dehydration Buffers: Dilute ethanol (EtOH) in distilled water to concentrations of 25, 40, 60, 80, 90, and 100% (*see* **Note 1**).
3. Fixing solution: 2.5% glutaraldehyde (stock): Dilute to 0.25% working concentration in PBS (*see* **Note 2**).
4. 10 mg/mL Acridine orange fluorescent stains: Dilute in 1× PBS to a working concentration of 20 μg/mL. Alternatively, 50 μg/mL Propidium Iodide (stock). Dilute in 1× PBS to a working concentration of 1 μg/mL (*see* **Note 3**).

2.2 Invasive E. coli Vector and Cell Lines

1. Bacterial growth media: Brain Heart Infusion (BHI) media. Prepare according to manufacturer instructions (*see* **Note 4**) and transfer to a 250 mL conical flask. Autoclave to sterilize. Store at room temperature.
2. Invasive *E. coli* vector: *E. coli* DH10B (pGB2Ωinv-hly, pEGFPN2) [2].
3. Fluorescent *E. coli* vector: *E. coli* DH10B (pGB2Ωinv-hly, pUltraRFP-KM) [2, 10].
4. Noninvasive control vector: *E. coli* DH10B (pEGFPN2) [2].
5. Cell lines: SK-N-SH (ATCC #:), HT1080 (ATCC #: CCL-121), GM02775 (Coriell Institute for Medical Research), or other desired cells that express β1 integrin (*see* **Note 5**).

2.3 Mammalian Cell Invasion

2.3.1 Cell Culture Media

1. Invasion media: RPMI or DMEM containing 10 mM L-glutamine but not containing serum. Store at 4 °C. Warm to 37 °C in a sterile water bath before us (*see* **Note 6**).
2. Complete culture media: RPMI or DMEM containing 10% FBS and 10 mM L-glutamine. Store at 4 °C. Warm to 37 °C in a water bath before use (*see* **Note 7**).
3. 0.05% Trypsin ETDA.

2.4 Cell Culture Materials

1. Six-well cell culture plates.
2. T-75 Cell culture flask.
3. 5 mL serological pipettes.
4. 10 mL serological pipettes.
5. Automatic pipettor.
6. Plate centrifuge.
7. Trypan blue.
8. Hemocytometer.
9. 15 mL Falcon centrifuge tubes.
10. 50 mL Falcon centrifuge tubes.
11. Temperature controlled water bath.
12. Silica gel (*see* **Note 8**).
13. Thin tipped forceps.

2.5 Microscopy

1. Hitachi SU8010 scanning electron microscope.
2. Nikon eclipse TE2000-E/C1*si*.
3. Glass microscope slides.
4. Plastic microscope coverslips.
5. Glass microscope coverslips.
6. Lint-free tissue.

3 Methods

3.1 Invasive E. coli Preparation

1. Spread *E. coli* on freshly prepared BHI agar and incubate at 37 °C overnight until single colonies are obtained.
2. Inoculate 10 mL of sterile BHI broth with freshly grown single colonies of invasive *E. coli* DH10B (pGB2Ωinv-hly, pUltraRFP-KM) and/or control noninvasive *E. coli* DH10B (pEGFPN2).
3. Grow *E. coli* cultures overnight in 10 mL of BHI broth in a sterile 50 mL Falcon tube at 37 °C with shaking at 220 rpm until late log/stationary phase is achieved approximately 18 h post inoculation or until OD_{600} reaches approximately 1.4.
4. Harvest bacteria by centrifugation at 6000 × *g* for 7 min at room temperature. Discard the spent bacterial media by gently pouring out the supernatant.
5. Resuspend the bacterial pellet in 0.9 mL of invasion media and mix thoroughly by pipetting to ensure the bacterial pellet is fully dissociated and evenly mixed.

3.2 Mammalian Cell Growth

1. Transfer a vial of mammalian cells from cryopreservation and thaw rapidly in a 37 °C water bath.
2. Using a 5 mL serological pipette, gently remove the thawed cells from the cryopreservation vial into a 15 mL Falcon tube containing 10 mL complete media and pipette gently to mix the cells. Centrifuge briefly at 200 × *g* for 5 min, remove the supernatant, and resuspend the pellet gently in 5 mL complete media before transferring to a T-75 cell culture flask containing 15 mL prewarmed complete cell culture media.
3. Incubate the cells in a 37 °C incubator with 5% CO_2 with humidity overnight.
4. Aspirate spent media from the culture vessel using a serological pipette. Wash cells twice using prewarmed 1× PBS by gently pipetting onto the walls of the culture vessel and gently swirling to ensure proper washing.
5. Remove 1× PBS using a serological pipette and add 15 mL of complete culture media to the culture flask. Incubate cells overnight at 37 °C with 5% CO_2 or grow cells in complete culture media until they reach approximately 70% confluent.
6. Aspirate spent media from the culture vessel using a serological pipette and wash cells twice using 1× PBS.
7. Aspirate 1× PBS from the culture flask and transfer 4 mL of prewarmed 0.05% trypsin to detach cells from the culture flask. Incubate cells in trypsin solution at 37 °C with 5% CO_2 and humidity for approximately 3–5 min or until cells are completely detached.

8. Gently add 5 mL of complete media to the culture vessel using a serological pipette to deactivate trypsin.
9. Collect detached cells into a 15 mL Falcon tube using a serological pipette. Centrifuge the cells at 225 × *g* for 5 min at room temperature.
10. Gently aspirate the supernatant using a serological pipette.
11. Resuspend the cell pellet in 1 mL of complete cell culture media. Pipette gently to ensure homogenous resuspension. Do not vortex.
12. Remove 10 μL of cells onto a hemocytometer using a micropipette and mix with 10 μL trypan blue dye to obtain a viable cell count.
13. Prepare cells for invasion by seeding HT1080, SK-N-SH, or GM02775 cells in 6-well plates containing plastic or glass coverslips at a density of 1×10^5 cells per well 24 h before invasion experiments (*see* **Note 9**).
14. On the day of invasion, remove growth media from the cell culture vessel using a sterile serological pipette and rinse cells twice with 1× PBS that has been prewarmed to 37 °C using a sterile water bath.
15. Remove 1× PBS from culture vessel using a serological pipette immediately before invasion.

3.3 Invasion Procedure

1. Place sterile plastic or glass coverslips at the bottom of a six-well plate. Seed cells into wells 24 h prior to invasion experiments by dispensing cells resuspended in complete culture media into wells using a serological pipette. Incubate at 37 °C with 5% CO_2 overnight (*see* **Note 9**).
2. On the day of invasion, discard spent culture media by aspiration using a serological pipette. Wash wells with 3 mL of 1× PBS by gently pipetting onto the side of wells (*see* **Note 10**).
3. Transfer bacterial cultures resuspended in invasion media onto cells grown on microscope coverslips in six-well plates. Centrifuge the plates at 225 × *g* for 10 min. Transfer the cells to a 37 °C in a CO_2 incubator for 1 h (Narayanan et al. 2013).
4. Remove bacteria that are not attached to cell membranes by washing wells three times with 3 mL of 1× PBS using a serological pipette (*see* **Note 11**).

3.4 Sample Preparation for Scanning Electron Microscopy

1. Fix cells by adding 2 mL of fixation buffer to each well using a serological pipette and incubate plates at 4 °C overnight (*see* **Note 12**).
2. Aspirate fixing solution using a serological pipette. Wash the wells and coverslips three times with 1× PBS that has been

prechilled at 4 °C. Swirl gently to ensure proper washing and aspirate wash buffer from the wells using a serological pipette (*see* **Note 13**).

3. Gently add 2 mL of 25% EtOH to the side of each well. Tilt the plate gently to ensure proper mixing and incubate at room temperature for 15 min.
4. Remove the 25% EtOH solution from wells and gently add in 40% EtOH. Repeat for each subsequent EtOH concentration until 100% (*see* **Note 14**).
5. Use forceps to remove coverslips from the well and place face up on lint-free tissue to remove excess liquid.
6. Air dry the coverslips for 5 min at room temperature and transfer them to a desiccator or a container with silica beads.
7. Mount coverslips on the SEM stage using double-sided carbon tape and sputter coat appropriately for viewing under the microscope.

3.5 Sample Preparation for Confocal Microscopy

1. Following invasion (*see* Subheading 3.3), add 3 mL of complete cell culture media containing 80 μg/mL of gentamicin to each well. Incubate at 37 °C in a CO_2 incubator for 1 h (*see* **Note 15**).
2. Aspirate media containing gentamicin using a serological pipette and replace with 3 mL complete culture medium containing 20 μg/mL gentamicin to each well. Incubate until the desired timepoint (*see* **Note 15**).
3. Aspirate culture medium and wash the cells three times with 3 mL of 1× PBS per well. Gently swirl the plates to ensure proper washing (*see* **Note 16**).
4. Add 3 mL of fixing solution and incubate at 4 °C overnight. (*see* **Note 12**).
5. Aspirate fixing solution and rinse wells with 3 mL of 1× PBS chilled to 4 °C (*see* **Note 17**).
6. Add 1 mL of fresh 1× PBS to each well. Using a micropipette, add acridine orange to each well. Swirl the plates gently to mix and store at 4 °C for 10 min.
7. Remove fluorescent stain by aspiration using a serological pipette. Gently rinse each well three times with 1× PBS (*see* **Note 18**).
8. Using forceps, remove microscope slides from each well and place face-up on lint-free tissue to dry for 10 min at room temperature.
9. Invert and mount coverslips onto glass slides for viewing under the microscope (*see* **Note 19**).

4 Notes

1. Use molecular grade ethanol. Gradient solutions of EtOH can be prepared in 50 mL centrifuge tubes to provide sufficient volume to allow sample dehydration in multiple six-well plates. Larger volumes can be prepared provided measurements are kept precise. It is advisable that ethanol gradient solutions are prepared fresh for each experiment.
2. Alternatively, 4% Paraformaldehyde (PFA) (stock) can be used. Dissolve 5 g of PFA into 100 mL of 1× PBS heated to 60 °C with stirring. Filter the solution when cooled. Dilute to 0.4% working concentration in PBS. Store at −20 °C. For rapid preparation of invaded adherent cell populations, fixation with 2.5% glutaraldehyde for 1 h at 37 °C provides the most stable results. However, rapid fixation may result in damage to specimen including shrinkage of bacteria and or cracks on cell membranes. Cells fixed with 0. 25% glutaraldehyde or 0.4% PFA overnight at 4 °C will result in better preserved cell specimen with more intact membranes. Glutaraldehyde and paraformaldehyde should be handled in a fume cabinet.
3. Application of acridine orange results in a fluorescent green stain to membrane, cytoplasmic components, and intracellular DNA. This should be considered when bacterial vectors constitutively expressing fluorescent proteins are used. Acridine orange may be substituted with propidium iodide to stain the same cellular components fluorescent red. Bacterial vectors not expressing alternate fluorescent proteins such as GFP or RFP will non-discriminatorily be stained by both fluorescent dyes.
4. Reconstituted BHI growth media should always be sterilized. Media can be prepared and stored for future experiments but is recommended to be made fresh.
5. Human Fabry disease fibroblasts (GM02775) are typically cultured in MEM containing Earle's salts and non-essential amino acids supplemented with 15% FBS. They can tolerate invasion media without serum for about 2 h before they begin to shrink and detach from cell culture vessels. Cells that express integrin are more permissive to attachment of invasive *E. coli*.
6. The base growth media used for each experiment is cell line specific. Most immortalized cell lines are maintained in RPMI or DMEM. It is important that media used for invasion does not contain serum as this leads to clumping of the bacterial vectors, which drastically reduces their attachment and invasion capabilities.
7. Complete cell culture media used to promote recovery of cells following the invasion process should contain all necessary components including 2 mM L-glutamine and 10% FBS. It is

helpful to increase the serum concentration to 12% following the invasion process to improve recovery rates and limit cell death.

8. We use silica gel to create transient desiccators to store and transport processed coverslips containing dehydrated samples. One teaspoon of silica gel can be placed in each well of a six-well plate and coverslips can be rested on the gel beads. Plates can be covered and air-sealed using parafilm to preserve dehydrated samples for SEM.
9. When using plastic coverslips, slightly bending a corner of the slips will enable much easier handling using forceps. Glass coverslips are ideal for confocal microscopy, but do not easily allow cells to attach. Higher cell numbers may be seeded to provide the desired confluence of approximately 60–70% on the day of invasion. We sterilize plastic and glass coverslips using 70% EtOH, followed by UV irradiation for 30 min prior to placement into sterile six-well plates.
10. Directly pipetting wash buffer onto coverslips may result in unwanted cell detachment. This is more of a concern with glass coverslips where cellular attachment to the coverslip is not as robust as it is to plastic. Washing steps can be repeated to ensure all traces of FBS are removed from the wells. Prolonged exposure of cells to PBS will however affect their morphology and may negatively impact the bacterial vectors ability to attach during invasion.
11. Plates can be quickly viewed under a standard light microscope to ensure that washing is thorough. Additional wash steps can be performed to ensure that bacteria are completely removed from areas of the culture surface that are not occupied by cells. Alternatively, coverslips can be picked up with forceps and gently dipped into 1× PBS in a sufficiently deep container. Multiple "wash basins" containing 1× PBS can be set up for the experiment to dip the coverslips in. It is however more scalable to wash coverslips inside six-well plates using a serological pipette.
12. Samples can be fixed with stock concentrations of glutaraldehyde or paraformaldehyde for 1 h at 37 °C, or alternatively with working concentrations at 4 °C overnight. Overnight fixation provides better sample quality by leaving cell and bacteria membranes intact. The quicker samples are fixed, the higher the likelihood of membrane shrinkage and cracking. We have found that fixing cells immediately following the invasion procedure allows the analysis of bacterial vector interaction with the cell membrane, prior to their internalization into the cytoplasm. Cells invaded at 37 °C for 1 h and fixed immediately after will feature bacteria external to the cell.

13. It is important to completely remove any traces of fixative as this might result in crystalline microstructures present in samples following dehydration. It is helpful to use a thin pipette tip to completely remove liquid in wells after each wash step following fixation.
14. For rapid procedures, each EtOH concentration can be used once. For complete dehydration, it is advisable to incubate samples in each EtOH concentration twice for 15 min each. For example, after 15 min of incubation in 25% EtOH, fresh 25% EtOH is used for a second incubation before the next concentration of 40% EtOH is used.
15. To analyze the internalization capability of bacteria, further incubation after the invasion process is required. We have observed that internalization of the invasive *E. coli* into the cytoplasm takes place between one- and three-hours post invasion. Complete culture media containing 80 μg/mL gentamicin can be used to eliminate external bacteria in the first hour following invasion. To observe internalization of *E. coli* at later timepoints, cells should be maintained in complete media containing 20 μg/mL gentamicin until their fixation timepoint.
16. Cells will attach more readily to plastic coverslips compared to glass coverslips. Care should be taken when rinsing cell monolayers growing on glass coverslips. Culture media, wash buffer, and fixative solutions should be added gently to the side of wells.
17. We have found that working with solutions chilled to 4 °C following sample fixation works best for preserving sample intactness. Transitioning between cold and warm conditions once the samples have been fixed increases the likelihood of damaging the samples by introducing cracks in the cell membranes, and by detaching bacterial cells that would be otherwise properly attached to the membrane.
18. Both propidium iodide and acridine orange bind strongly to DNA and passively to the cytoplasm and cell membrane. Cells can be washed extensively if the passive staining of the cytoplasm and cell membrane is overwhelming compared to DNA stained in the nucleus.
19. In place of specialized mounting reagents, a small drop of glycerol can be used as a mounting reagent. Lint-free tissue can be used to wipe away excess glycerol from the mounted microscope slide. Coverslips should be prevented from moving on the glass slide when mounted by gently dabbing excess glycerol instead of wiping.

Acknowledgments

This work was supported by a MOSTI eScience grant 02-02-10-SF0252 from the Ministry of Science, Technology and Innovation, Malaysia, to K.N. A.N.O. and A.W.R.N. were supported by Higher Degree Research Scholarships from Monash University Malaysia.

References

1. Celec P, Gardlik R (2017) Gene therapy using bacterial vectors. Front Biosci (Landmark Ed) 22:81–95
2. Narayanan K, Warburton PE (2003) DNA modification and functional delivery into human cells using *Escherichia coli* DH10B. Nucleic Acids Res 31(9):e51
3. Chang CH, Cheng WJ, Chen SY, Kao MC, Chiang CJ, Chao YP (2011) Engineering of *Escherichia coli* for targeted delivery of transgenes to HER2/neu-positive tumor cells. Biotechnol Bioeng 108(7):1662–1672. https://doi.org/10.1002/bit.23095
4. Barbier M, Damron FH (2016) Rainbow vectors for broad-range bacterial fluorescence labeling. PLoS One 11(3):e0146827. https://doi.org/10.1371/journal.pone.0146827
5. Dirks BP, Quinlan JJ (2014) Development of a modified gentamicin protection assay to investigate the interaction between campylobacter jejuni and *Acanthamoeba castellanii* ATCC 30010. Exp Parasitol 140:39–43. https://doi.org/10.1016/j.exppara.2014.03.012
6. VanCleave TT, Pulsifer AR, Connor MG, Warawa JM, Lawrenz MB (2017) Impact of gentamicin concentration and exposure time on intracellular Yersinia pestis. Front Cell Infect Microbiol 7:505. https://doi.org/10.3389/fcimb.2017.00505
7. Jones CH, Rane S, Patt E, Ravikrishnan A, Chen CK, Cheng C, Pfeifer BA (2013) Polymyxin B treatment improves bactofection efficacy and reduces cytotoxicity. Mol Pharm 10(11):4301–4308. https://doi.org/10.1021/mp4003927
8. Narayanan K, Lee CW, Radu A, Sim EU (2013) *Escherichia coli* bactofection using Lipofectamine. Anal Biochem 439(2):142–144. https://doi.org/10.1016/j.ab.2013.04.010
9. Osahor A, Deekonda K, Lee CW, Sim EU, Radu A, Narayanan K (2017) Rapid preparation of adherent mammalian cells for basic scanning electron microscopy (SEM) analysis. Anal Biochem 534:46–48. https://doi.org/10.1016/j.ab.2017.07.008
10. Mavridou DA, Gonzalez D, Clements A, Foster KR (2016) The pUltra plasmid series: a robust and flexible tool for fluorescent labeling of Enterobacteria. Plasmid 87-88:65–71. https://doi.org/10.1016/j.plasmid.2016.09.005
11. Damas-Souza DM, Nunes R, Carvalho HF (2019) An improved acridine orange staining of DNA/RNA. Acta Histochem 121(4):450–454. https://doi.org/10.1016/j.acthis.2019.03.010
12. Zhang N, Fan Y, Li C, Wang Q, Leksawasdi N, Li F, Wang S (2018) Cell permeability and nuclear DNA staining by propidium iodide in basidiomycetous yeasts. Appl Microbiol Biotechnol 102(9):4183–4191. https://doi.org/10.1007/s00253-018-8906-8

Part II

Nanoparticles

Chapter 4

Assessing Nucleic Acid: Cationic Nanoparticle Interaction for Gene Delivery

Moganavelli Singh

Abstract

The assessment of the efficient binding between a nucleic acid and its associated nanoparticle is crucial for gene delivery. Emerging from the extensive search for versatile gene carriers, are complexes formed between nucleic acids and nonviral nanocarriers that promise to be viable alternatives to the predominantly viral-based gene delivery vehicles. However, much is still to be known about the exact structure and physicochemical properties of such nanocomplexes. This chapter will concentrate on cationic lipid, polymer, and functionalized metal nanoparticles and their interaction with nucleic acids by direct conjugation or electrostatic interaction. Methods commonly employed to evaluate the nature and extent of nucleic acid interactions with cationic nanocarriers, such a nucleic acid binding, nuclease protection, and dye displacement assays will be described. In addition, the ultrastructural morphology, size, and zeta potential of these nanocomplexes, which are crucial for their cellular uptake and intracellular trafficking, will be assessed using electron microscopy, fluorescent detection, and nanoparticle tracking analysis (NTA). These assays have the ability to visualize and quantify the interaction and can also be used to complement each other, in addition to providing confirmation of the formation of the relevant nanocomplexes.

Key words Nucleic acids, Nanoparticles, Nanocarriers, Interaction, Complexes, Gene delivery

1 Introduction

The exhaustive search for safe and versatile gene delivery systems has produced an array of different systems to date, ranging from the popular viral systems to lipids, polymers, and inorganic nanoparticles such as gold, selenium, silver, iron oxide, and platinum, to name a few. Important to the use of these carrier systems are complexes formed between the nucleic acid and the nanocarrier. The exact structure and properties of such nanocomplexes is crucial to the eventual cellular uptake, intracellular trafficking, and expression or silencing of specific genes. Some of the more popular nonviral nanodelivery systems, include polymeric micelles, dendrimers, lipid and polymeric nanoparticles, metal nanoparticles, quantum dots (QDs), and liposomes [1].

Kumaran Narayanan (ed.), *Bio-Carrier Vectors: Methods and Protocols*, Methods in Molecular Biology, vol. 2211, https://doi.org/10.1007/978-1-0716-0943-9_4,

Cationic polymers, cationic polymer functionalized metal nanoparticles, cationic lipids, or cationic lipids assembled into lipid bilayer vesicles or liposomes, expose their positively charged amine group/s to the aqueous medium, which electrostatically bind to DNA and other nucleic acids such as small interfering RNA (siRNA) and messenger RNA (mRNA). These amine groups are the nucleic acid binding moieties, and upon electrostatic interaction, condense these carriers into smaller transportable units, termed lipoplexes for lipid-based complexes, or polyplexes for polymer-based complexes [2, 3]. Due to the small or nano-size of these complexes (including the inorganic nanoparticles), we shall refer to all such complexes as nanocomplexes.

Polymers such as dendrimers [4, 5], chitosan [6, 7], poly-L-lysine (PLL) [7, 8], and polyethyleneimine (PEI) [9, 10] have gained much interest due to their ability to bind and condense DNA and have hence been a popular choice for cationic functionalization of inorganic nanoparticles [11]. Hence, these interactions are mostly ionic in nature. Nanocomplexes can be dynamic in nature, and depending on the formulation conditions different conformations may arise. This chapter looks mostly at electrostatic formulations, with one of the first simplistic models of this type described by Felgner and Ringold in 1989 for the interaction of cationic liposomes with plasmid DNA [12], where a single plasmid is trapped within four liposomes. However, considering the different isoforms especially the supercoiled forms of the DNA, likely structures that can occur are those resembling the nanoparticles clustered like grapes around a more branched DNA molecule [13], or a "bead on a string" conformation [14], or a "spaghetti-meatball" type appearance [15]. In all cases, the cationic nanoparticles will continually bind to the DNA until a critical point (optimal binding ratio) is attained.

In the case of inorganic and metal nanoparticles, depending on the type of synthesis, they can adopt various shapes, such as rods, spheres, diamonds, stars, cubes, and triangles among others. Cellular uptake can be dependent on shape due to differences in the particle curvature, which ultimately influences the particle contact area with the cell membrane [16, 17]. While most studies have indicated that spherical nanoparticles have a higher propensity to be internalized in vitro in epithelial cells compared to rod-shaped particles of similar dimensions [18, 19], a contrasting study has shown that rod-shaped nanoparticles are more readily taken up by intestinal epithelial cells, in vitro, than spherical nanoparticles [17]. However, these two conformations seem most favorable and hence popular as nanodelivery vehicles. Hence, it is crucial to fully examine these formulations and their optimal binding prior to use in vitro or in vivo.

In this chapter, we explore some common and simple methods for the assessment of nanocomplex formation. Techniques utilizing fluorescence (dye displacement), agarose gel electrophoresis to determine DNA/RNA binding and protection from serum nucleases, transmission electron microscopy for a visual determination of the ultrastructural morphology of the nanocomplexes, and nanoparticle tracking and analysis to assess the effect of the interactions on the size, charge, and stability of the nanocomplex. These methods all play significant roles in the elucidation of the mass/charge ratios required for optimal nucleic acid:nanocarrier interaction, the structural features of the resulting nanocomplexes, their stability, and their ability to bind and enter cells. Hence, these preliminary studies will provide valuable information as to the potential of the nanocarriers for further gene therapeutic investigations.

2 Materials

All reagents, nanoparticles, and the nucleic acid stock solution can be made up in ultrapure water (preferably 18 MΩ) or sterile HEPES buffered saline (HBS) at physiological pH 7.4.

2.1 Molecular Biology Reagents and Materials

1. Nucleic Acid Stock: 0.25 or 0.5 μg/μL in ultrapure water or HBS.
2. Ethidium Bromide: 10 mg/mL in ultrapure water (*see* **Note 1**).
3. 5% Sodium Dodecyl Sulfate (SDS): 5 g SDS in 100 mL H_2O.
4. 0.5 M ethylenediamine tetra-acetic acid (EDTA), pH 8: 18.6 g EDTA in 100 mL distilled water (*see* **Note 2**).
5. Ultrapure Agarose (Molecular Biology Grade).
6. Fetal Bovine serum (FBS)/RNase A/DNase.
7. Flat-bottom black multiwell plate for fluorescence measurements.
8. Formvar/carbon-coated grids for electron microscopy.

2.2 Buffers and Solutions

1. HEPES Buffered Saline (HBS): 20 mM HEPES and 150 mM NaCl, pH 7.4.
2. Gel Loading Buffer: 25 mg bromophenol blue, 25 mg xylene cyanol, 4 g sucrose in 10 mL H_2O (*see* **Note 3**).
3. 10× Electrophoresis Buffer: 0.36 M Tris–HCl, 0.3 M NaH_2PO_4, 0.1 M EDTA, pH 7.4–7.5 (*see* **Note 4**).

2.3 Equipment

1. Electrophoresis apparatus (mini-sub type) with power supply.
2. Gel Documentation System/Gel Imager.

3. Spectrofluorometer.
4. Transmission Electron Microscope (TEM).
5. Nanoparticle analyzer, e.g., Nanosight NS500.

3 Methods

All procedures can be conducted on the bench top at room temperature, unless otherwise required.

3.1 Agarose Gels (1%)

1. Suspend 0.2 g agarose in 18 mL ultrapure H_2O and boil to dissolve (*see* **Note 5**).
2. Cool to approximately 75 °C and then add 2 mL of 10× electrophoresis buffer.
3. At this point, ethidium bromide (1 μg/mL) can be included in the gel.
4. Place gel tray into a gel-casting tray. If no casting tray is available, then seal the gel tray with tape.
5. Add a multiwell comb about 1 cm away from the end of the tray.
6. Once the gel has cooled to 60–62 °C, pour into the gel tray and allow gel to set for 1 h or less.

3.2 Electrophoretic Mobility Shift

1. Use the 1% (for DNA) or 2% (for RNA) agarose gel as outlined in Subheading 3.1.
2. While the gel is setting, prepare the nanocomplexes by using varying amounts of the cationic nanoparticle and a constant amount of the DNA/RNA (0.25 or 0.5 μg/μL) (*see* **Note 6**).
3. Incubate the nanocomplexes at room temperature for 20–30 min. However, for some functionalized or cationized inorganic nanoparticle-based carriers such as gold, an incubation period of 1 h may be required [20].
4. Following the incubation period, add 1–2 μL of the gel loading buffer to the nanocomplexes.
5. Load the nanocomplexes into the respective wells in the agarose gel. Include a naked nucleic acid control in one well (*see* **Note 7**).
6. Place the loaded gel into an electrophoresis apparatus containing 1× electrophoresis buffer (*see* **Note 8**).
7. Conduct the electrophoresis at 50 V for 1 h for DNA [21] and for 30 min for RNA [22] molecules.
8. Remove the gel from the apparatus and view and obtain images under UV transillumination at 300 nm (*see* **Note 9**).
9. A typical gel result for DNA is depicted in Fig. 1 (*see* **Note 10**).

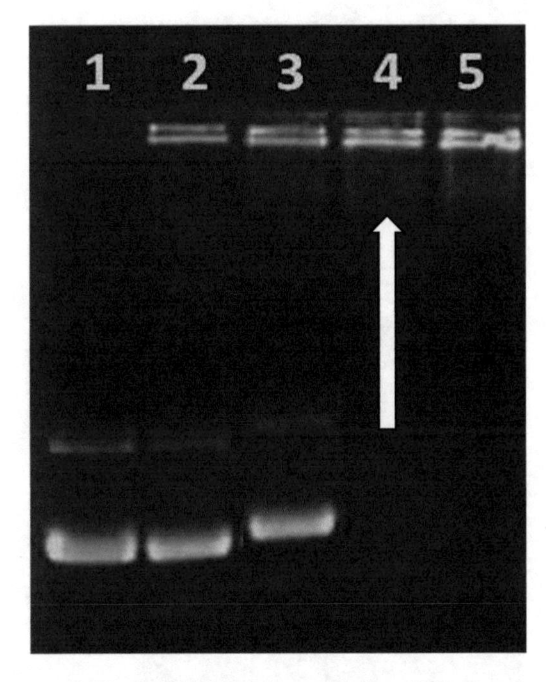

Fig. 1 Agarose gel electrophoresis of nanocarrier : DNA complexes. Lane 1: naked DNA, lanes 2–5: nanocarrier:DNA complexes of varying ratios. Brightly stained bands in the wells reflect the formation of electroneutral complexes. Arrow indicates endpoint or optimal ratio

3.3 Ethidium Bromide Displacement

1. Set a spectrofluorometer for excitation and emission wavelengths of 520 nm and 600 nm, respectively.
2. Add approximately 100 μL of HBS and 2 μL of ethidium bromide into separate wells in a 96-well flat-bottom black microplate and mix thoroughly. Single cuvettes can be used depending on the instrument utilized.
3. For RNA, SYBR Green II [23] can be used instead of ethidium bromide if greater sensitivity is required (*see* **Note 11**).
4. Measure the fluorescence and set as 0% baseline relative fluorescence.
5. Thereafter, add small increments of the nanocarrier (same range used in Subheading 3.2). Measure fluorescence after each addition, until a point of inflection in fluorescence is attained (*see* **Notes 12** and **13**).
6. Plot the relative fluorescence against the mass (μg) of the nanocarrier (*see* **Note 14**). A typical graph is illustrated in Fig. 2. This assay is also known as an ethidium bromide intercalation assay.

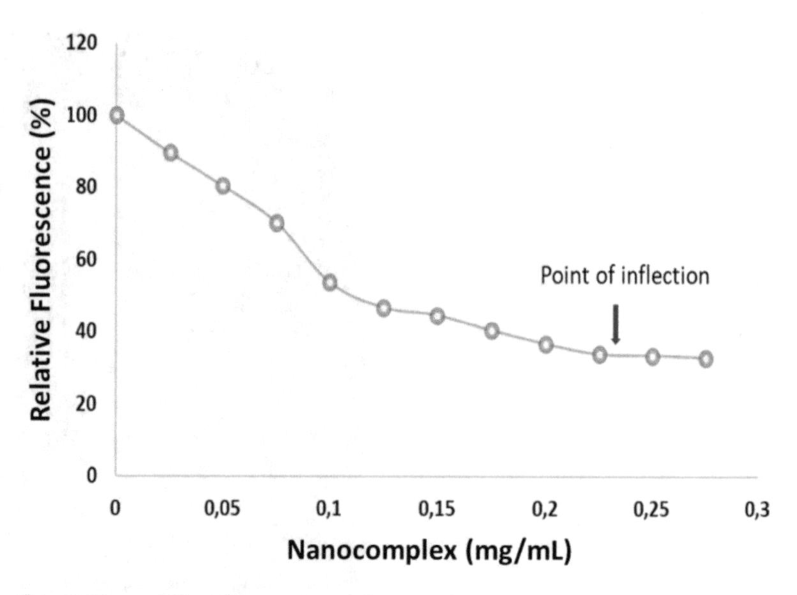

Fig. 2 The relative fluorescence of complexes using the ethidium bromide intercalation assay

3.4 Enzyme Protection Assay

Based on the results obtained from Subheading 3.2, generally three ratios are selected viz., sub-optimum, optimum, and supra-optimum binding ratios for the nanocomplexes, and are used for the protection assay (*see* **Note 15**).

1. Make up the nanocomplexes to a final constant volume with HBS. We commonly use 10 μL as the final volume as this is sufficient and will not overload the wells in the prepared gel.
2. Incubate the mixtures at room temperature for 30–60 min to allow for nanocomplex formation.
3. Thereafter, add fetal bovine serum (FBS) to the nanocomplexes to a final concentration of 10% (*see* **Note 16**).
4. Set up two controls: positive control containing naked DNA or RNA (0.25–0.5 μg/μL), and a negative control containing naked DNA or RNA (0.25–0.5 μg/μL) treated with 10% FBS.
5. To facilitate nuclease digestion, all samples containing FBS must be incubated at 37 °C for 4 h (*see* **Note 17**).
6. Terminate the reaction by the adding EDTA to all samples to a final concentration of 10 mM EDTA (*see* **Note 18**).
7. Thereafter, add 5% sodium dodecyl sulfate (SDS) to all samples to a final concentration of 0.5% and incubate the samples at 55 °C for 20 min (*see* **Note 19**).
8. Subject the samples to agarose gel electrophoresis for 30 min and view as previously described in Subheading 3.2. A typical nuclease digestion of an mRNA sample is shown in Fig. 3.

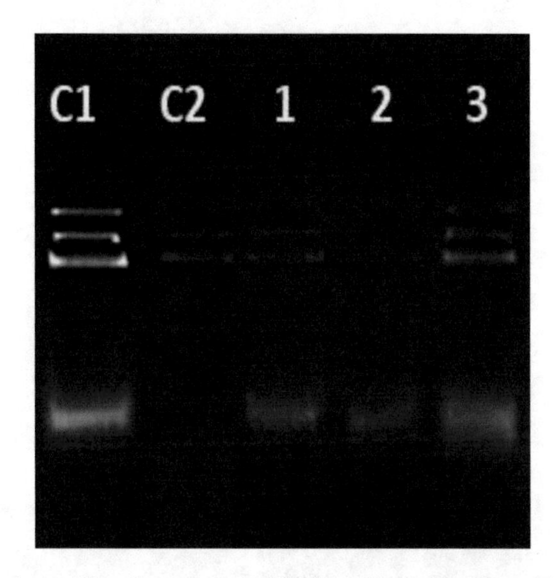

Fig. 3 Nuclease digestion of an mRNA-based nanocomplex. C1 contained undigested mRNA and C2 contained FBS-digested mRNA. Lane 1 = nanocomplex at sub-optimum binding ratio, lane 2 = nanocomplex at optimum binding ratio, and lane 3 = nanocomplex at the supra-optimum binding ratio

3.5 Electron Microscopy (EM)

Examine nanocomplexes at the desired or the optimum ratios obtained from the electrophoretic mobility shift assay (Subheading 3.2). Nanocomplexes must be prepared fresh on the day.

1. Add small amounts (1–2 μL) of the nanocomplex to formvar/carbon-coated grids.
2. Air dry the grids and view sample under TEM (*see* **Notes 20** and **21**).
3. Once grids are dry, place the samples in liquid nitrogen for a few minutes and then view if using Cryo-TEM [27].

3.6 Nanoparticle Tracking Analysis (NTA)

1. NTA is a robust method used to accurately determine the size, dispersity, zeta potential, and colloidal stability in real time [24] (*see* **Notes 22–24**).
2. Dilute nanocomplexes depending on the population density of your nanocomplexes, to up to 1:1000 in ultrapure water so as to produce particle numbers appropriate to the sensitivity limits of the instrument (e.g., Nanosight NS500).
3. Prime the system, flush with ultrapure water, and set the camera at the zero position prior to loading of samples.
4. Visualize particles using the focused laser beam (405 nm, 60 mW) and the on-board sCMOS camera.

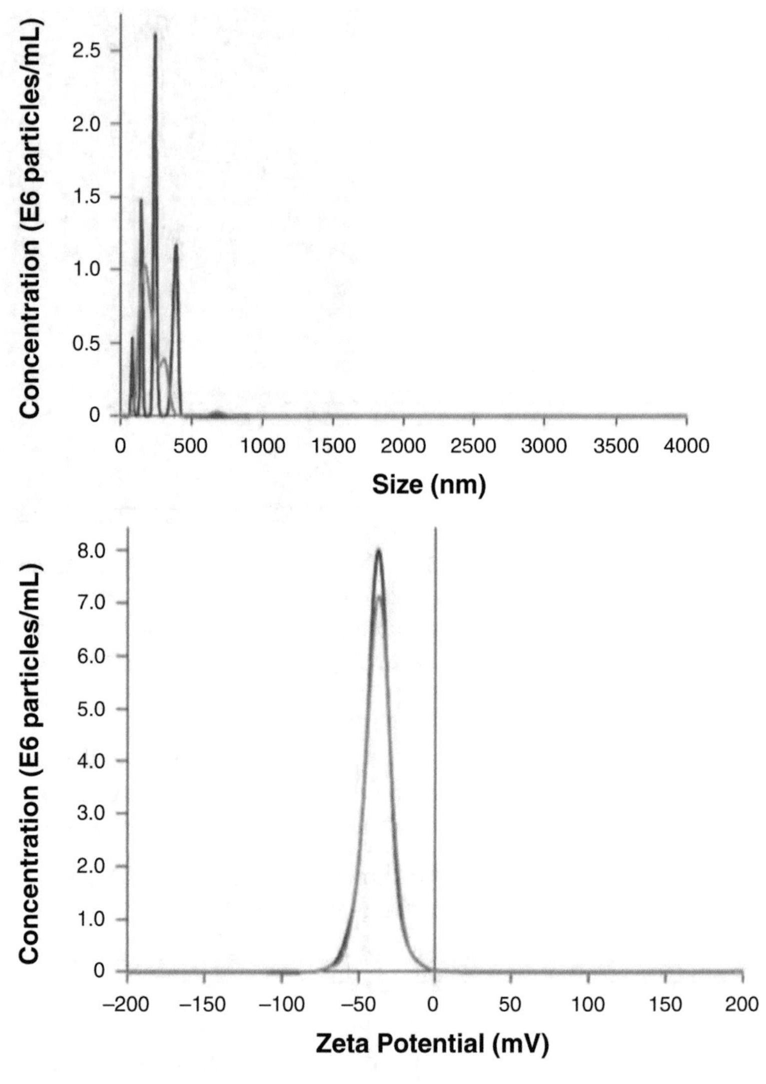

Fig. 4 NTA profile showing the size distribution and zeta potential of a siRNA: functionalized gold nanocomplex

5. Capture videos of the particle motion, usually at 60 s durations and at a frame rate of 30 frames/s within the chamber upon the application of 10 V, under both positive and negative polarity. Maintain temperature at 25 °C during measurements, with an assumed sample viscosity of 0.9 cP and dielectric constant of 80.0. A typical result obtained from NTA can be seen in Figs. 4 and 5. *See* **Note 25** for discussion about differences that may occur with respect to nanocomplex sizes using TEM and NTA.

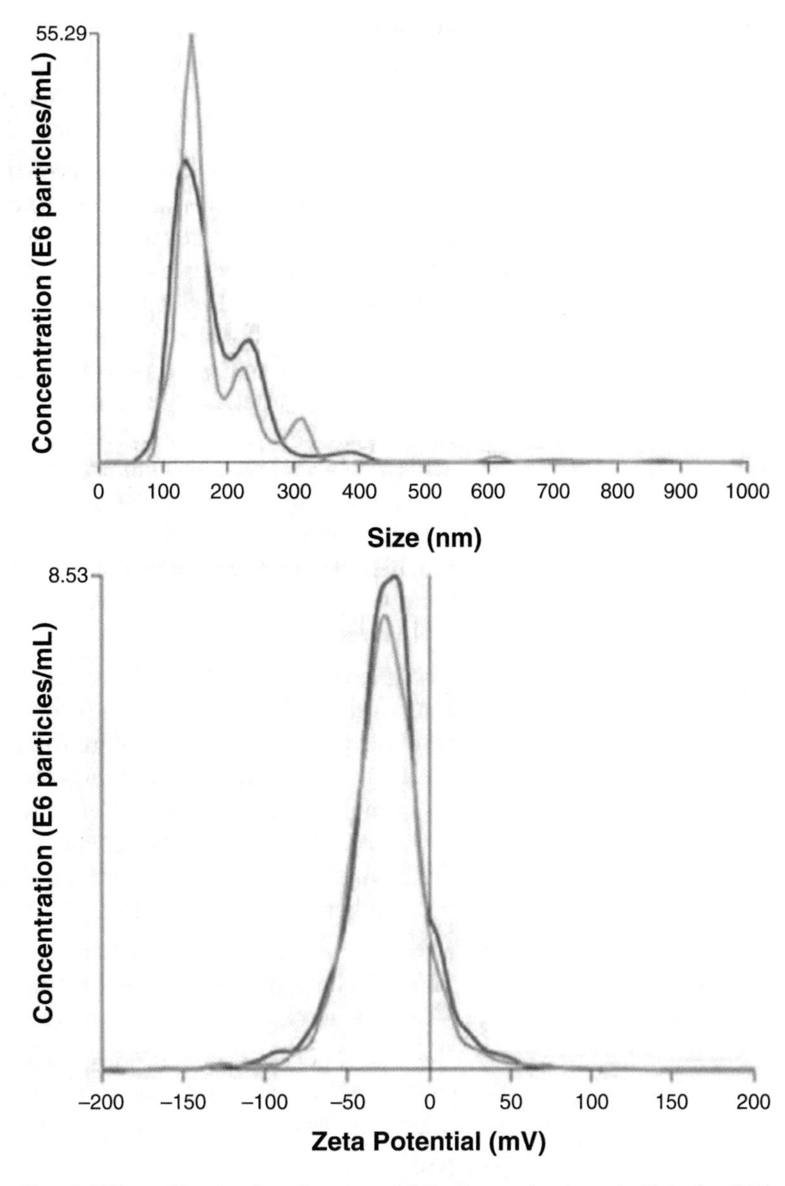

Fig. 5 NTA profile showing the size distribution and zeta potential of a DNA: liposome nanocomplex

4 Notes

1. Ethidium bromide in solution can be purchased commercially.
2. To facilitate better solubility, adjust pH to 8.0 with NaOH.
3. The gel loading buffer can also be prepared by adding 25 mg bromophenol blue and 4 g sucrose; or 25 mg bromophenol blue, 25 mg xylene cyanol, and 3 mL glycerol in 10 mL H_2O.
4. Other buffers such a Tris-borate-EDTA (TBE) can also be used as effectively if available in your laboratory.

5. A 1% gel can be used for DNA, but for RNA a 2% gel may be required [22]. Double the quantities required for the agarose in this case. If one desires a thicker gel, the volume can be doubled, but do not forget to double all other amounts as well.
6. Preparation of nanocomplexes ultimately will depend on the binding of the cationic nanocarrier to the nucleic acid, which in turn depends on the relative density of positive and negative charges on the respective components that will facilitate electrostatic interaction. Metallic and inorganic nanoparticles are commonly functionalized with polymers such as chitosan, poly-L-lysine, dendrimers, polyethyleneimine (PEI) which confer positive charges for nanocomplexation. When deciding on ratios it is wise to start from at least a low (nanoparticle:nucleic acid) 1:1 (w/w) ratio and increase the nanoparticle amounts by at least 0.5 μg each time. Cationic polymers on their own will naturally have lower binding ratios so expect to use less of the polymer to bind the nucleic acid. Often mRNA and siRNA nanocomplexes would have higher binding ratios than for DNA-based nanocomplexes.
7. The naked DNA is used as it would migrate freely into the gel and produce the expected bands for the supercoiled, circular, and linear forms. Since the migration of this DNA will be used as a control against which the migration of unbound DNA will be compared, the naked DNA used must be of the same concentration and of the same type, e.g., substituting a plasmid DNA (e.g., pBR322 DNA) with calf thymus DNA as the naked control, may skew the results.
8. For most mini-sub electrophoresis apparatus approximately 200–250 mL of 1× electrophoresis buffer is needed. This should be diluted from the 10× electrophoresis buffer stock.
9. Dedicated gel documentation or imaging systems are available where one can view and capture images. Do not view UV light directly as it is harmful to the eyes. Wear UV protective glasses.
10. Nucleic acid that binds to the carrier will be retarded in the gel and appear closer to the wells, while the DNA/RNA on its own will migrate into the gel. When all the negative charges on the nucleic acid are completely bound to the positive charges of the nanocarrier, an electroneutral complex is produced, that remains in the well (Fig. 1, lanes 4 and 5). This is evident by the bright fluorescence seen in the wells. This is regarded as the optimum binding ratio (lane 4). The ratio above and ratio below this are referred to as the supra-optimum (lane 5) and sub-optimum (lane 3) ratios, respectively. We usually use these three ratios for all subsequent evaluations in vitro. This assay is also known as the band shift or gel retardation assay.

11. This assay can also be conducted by substituting SYBR Green II RNA stain for ethidium bromide when looking at RNA binding and condensation [23]. We dilute the stock (1:10,000) in HBS, and use approximately 202 μL to establish a baseline fluorescence. This amount can be optimized according to your ratios obtained from the electrophoretic mobility shift assay. Fluorescence intensities upon introduction of siRNA and carrier are measured at excitation and emission wavelengths of 497 nm and 520 nm, respectively. Results are reported as described in Subheading 3.3.
12. Generally 1 μL aliquots of the nanocomplexes are suitable, but higher volumes can be used depending on the stock concentration. Our nanocarrier stock is usually around 0.025 μg/μL.
13. This assay measures the level of condensation of the nanocomplexes, which is essential, particularly at the point of endosomal escape for the release and delivery of the nucleic acid therapeutic agents [25].
14. Fluorescence measurements are calculated as relative fluorescence (Fr) using the equation: Fr (%) $= (F_i - F_0)/(F_{max} - F_0) \times 100$, where F_0 is the fluorescence intensity of EtBr/HBS mixtures, F_i is the fluorescence intensity at each concentration of the nanocarrier, and F_{max} is the fluorescence intensity of the ethidium bromide/nucleic acid mixture.
15. The extent of protection offered by the nanocarrier to the nucleic acid in an environment that imitates an in vivo system can be assessed using a nuclease protection assay. This assay will provide an indication as to the potential stability of the complex when administered in vivo.
16. For a more stringent testing of protection afforded by a nanocarrier to both DNA and RNA, the use of the enzymes DNase and RNase A [22] can be substituted for FBS.
17. Incubation at 37 °C is essential for optimum enzyme activity and will mimic an in vivo system.
18. EDTA is added to stop the action of the enzyme.
19. The use of SDS is needed to reduce the electrostatic interaction between the nanocarrier and the nucleic acid. This will release the nucleic acid which will the migrate into the gel as observed in Fig. 3. However, inorganic nanoparticles often show resistance in releasing the nucleic acid under these conditions and may require the use of higher SDS concentrations, or the use of heparin sulfate (an expensive option). Hence, for these nanocarriers, bright fluorescing bands are often visualized in the wells indicating that the nucleic acid was still bound to the nanocarrier and not released. Nevertheless, this also suggests protection of the nucleic acid by the nanocarrier.

20. Transmission electron microscopy (TEM) is commonly used to view the morphology and establish homogeneity and size distribution of nanocomplexes. Scanning electron microscopy (SEM) and High Resolution Transmission Electron Microscopy (HRTEM) can also be employed, but from our experience they have been more suitable to view the nanocarriers on their own.
21. Staining with 0.5% uranyl acetate for lipid-based nanocomplexes may be required as they are not as dense as metal nanoparticles [26]. Cryo-TEM is also another option that works well for most nanocomplexes.
22. The Stokes-Einstein equation was used to determine particle size distribution based on tracking the trajectories of particles in Brownian motion within the laser scatter volume [24].
23. Zeta potentials is calculated from the mean electrophoretic mobility using Smoluchowski approximation which is based on Laser-Doppler microelectrophoresis [24]. Particle concentrations are estimated based on the number of particles in the field of view in relation to the scattering volume. These calculations generally are done by the associated software of the NTA instrument.
24. The zeta potential values greater than +30 mV or less than −30 mV are indicative of excellent colloidal stability [28]. Note that Dynamic light Scattering (DLS) can also be used to obtain size and zeta potential.
25. The sizes of nanocomplexes from TEM and NTA often differ. This can be due to the fact that TEM measures the samples in a dry state while NTA measures samples in an aqueous medium [29]. NTA sizes may be closer to that in an *in vivo* system.

Acknowledgments

Research conducted in this area was partly funded by the National Research Foundation of South Africa (Grant ID 81289). The postgraduate students of the Nano-Gene and Drug Delivery Group are acknowledged for their contribution in optimizing the protocols described.

References

1. Conti M, Tazzari V, Baccini C, Pertici G, Serino LP, De Giorgi U (2006) Anticancer drug delivery with nanoparticles. In Vivo 20:697–702
2. Brown MD, Schätzlein AG, Uchegbu IF (2001) Gene delivery with synthetic (non-viral) carriers. Int J Pharm 229:1–21
3. Felgner PL, Gadek TR, Holm M, Roman R, Chan HW, Wenz M, Northrop JP, Ringold GM, Danielson M (1987) Lipofection: a highly efficient lipid-mediated DNA-transfection procedure. Proc Natl Acad Sci U S A 84:7413–7417

4. Mbatha LS, Maiyo FC, Singh M (2019) Dendrimer functionalized folate-targeted gold nanoparticles for luciferase gene silencing in vitro: a proof of principle study. Acta Pharm 69(1):49–61
5. Mbatha LS, Singh M (2019) Starburst poly (amidoamine) dendrimer grafted gold nanoparticles as a scaffold for folic acid-targeted plasmid DNA delivery in vitro. J Nanosci Nanotechnol 19(4):1959–1970
6. Corsi K, Chellat F, Yahia L, Fernandes JC (2003) Mesenchymal stem cells, MG63 and HEK293 transfection using chitosan-DNA nanoparticles. Biomaterials 24:1255–1264
7. Lazarus G, Singh M (2016) Cationic modified gold nanoparticles show enhanced gene delivery in vitro. Nanotechnol Rev 5(5):425–434
8. Wu GY, Wu CH (1987) Receptor-mediated in vitro gene transformation by a soluble DNA carrier system. J Biol Chem 262:4429–4432
9. Boussif O, Lezoulach F, Zanto MA, Mergny MP, Scherman D, Demeniex B, Behr JP (1995) A versatile vector for gene and oligonucleotide delivery into cells in culture and in vivo: polyethyleneimine. Proc Natl Acad Sci U S A 92:7297–7314
10. Lazarus G, Singh M (2016) In vitro cytotoxic activity and transfection efficiency of polyethyleneimine functionalized gold nanoparticles. Colloids Surf B Biointerfaces 145:906–911
11. Lazarus GG, Revaprasadu N, Jopez-Viota J, Singh M (2014) The electrokinetic characterization of gold nanoparticles, functionalized with cationic functional groups, and its' interaction with DNA. Colloids Surf B Biointerfaces 121:425–431
12. Felgner PL, Ringold GM (1989) Cationic liposome mediated transfection. Nature 337:387–388
13. Boles TC, White JH, Cozzarelli NR (1990) Structure of plectonemically supercoiled DNA. J Mol Biol 213:931–951
14. Gershon H, Ghirlando R, Guttmann SB, Minsky A (1993) Mode of formation and structural features of DNA-cationic liposome complexes used for transfection. Biochemistry 32:7143–7151
15. Sternberg B, Sorgi FL, Huang L (1994) New structures in complex formation between DNA and cationic liposomes visualized by freeze-fracture electron microscopy. FEBS Lett 356 (23):361–366
16. Albanese A, Tang PS, Chan WC (2012) The effect of nanoparticle size, shape, and surface chemistry on biological systems. Annu Rev Biomed Eng 14:1–16
17. Banerjee A, Qi J, Gogoi R, Wong J, Mitragotri S (2016) Role of nanoparticle size, shape and surface chemistry in oral drug delivery. J Control Release 238:176–185
18. Caldorera-Moore M, Guimard N, Shi L, Roy K (2010) Designer nanoparticles: incorporating size, shape and triggered release into nanoscale drug carriers. Expert Opin Drug Deliv 7 (4):479–495
19. Hoshyar N, Gray S, Han H, Bao G (2016) The effect of nanoparticle size on in vivo pharmacokinetics and cellular interaction. Nanomedicine 11(6):673–692
20. Akinyelu J, Singh M (2018) Chitosan stabilized gold-folate-poly(lactide-co-glycolide) nanoplexes facilitate efficient gene delivery in hepatic and breast cancer cells. J Nanosci Nanotechnol 18(7):4478–4486
21. Singh M, Ariatti M (2003) Targeted gene delivery into HepG2 cells using complexes containing DNA, cationized asialoorosomucoid and activated cationic liposomes. J Control Release 92(3):383–394
22. Daniels AN, Singh M (2019) Sterically stabilised siRNA:gold nanocomplexes enhance c-MYC silencing in a breast cancer cell model. Nanomedicine 14(11):1387–1401
23. Dorasamy S, Singh M, Ariatti M (2009) Rapid and sensitive fluorometric analysis of novel galactosylated cationic liposome interaction with siRNA. Afr J Pharm Pharmacol 3 (12):635–638
24. Malloy A (2011) Count, size and visualize nanoparticles. Mater Today 14(4):170–173
25. Lasic DD, Strey H, Stuart MCA, Podgornik R, Frederik PM (1997) The structure of DNA-liposome complexes. J Am Chem Soc 119:832–833
26. Singh M, Ariatti M (2006) A cationic cytofectin with long spacer mediates favourable transfection in transformed human epithelial cells. Int J Pharm 309:189–198
27. Gorle S, Ariatti M, Singh M (2014) Novel serum-tolerant lipoplexes target the folate receptor efficiently. Eur J Pharm Sci 59 (1):83–93
28. Gibson N, Shenderova O, Luo TJM, Moseenkov S, Bondar V, Puzyr A, Purtov K, Fitzgerald Z, Brenner DW (2009) Colloidal stability of modified nanodiamond particles. Diamond Relat Mater 18(4):620–626
29. Akinyelu J, Singh M (2019) Folate-tagged chitosan functionalized gold nanoparticles for enhanced delivery of 5-fluorouracil to cancer cells. App Nanosci 9(1):7–17

Chapter 5

Catanionic Hybrid Lipid Nanovesicles for Improved Bioavailability and Efficacy of Chemotherapeutic Drugs

Xuemei Xie, Dan He, Yan Wu, Tingting Wang, Cailing Zhong, and Jingqing Zhang

Abstract

Catanionic nanovesicles are attractive as a novel class of delivery vehicle because they can increase the stability, adsorption, and cellular uptake of a broad range of drugs. These hybrid lipid nanocarriers consist of solid and liquid lipids, which are biocompatible and biodegradable. Since liquid lipid is added to the nanocarrier, the lipids are present in a crystalline defect or amorphous structure state. As a result, hybrid lipid nanocarriers have a higher drug loading capability and suffer less drug leakage during preparation and storage compared to the pure lipid nanocarriers. Catanionic nanovesicles have been shown to increase stability, adsorption, cellular uptake, apoptosis induction, tumor cell cytotoxicity, and antitumorigenic effect, making it a highly desirable vehicle for drug delivery. For example, the anticancer compound curcumin (CC) have shown great promise to cure cancers such as lung cancer, breast cancer, stomach cancer, and colon cancer. However, like many potential antitumor drugs, CC on its own has poor water solubility, easy photodegradation, chemical instability, low bioavailability, rapid metabolism, and fast systematic clearance, which severely limits its clinical applications. In this chapter, we demonstrate the use of catanionic nanovesicles to improve the bioavailability and efficacy of CC for anticancer applications. This technique can be easily adapted for delivery and evaluation of other bioactive compounds.

Key words Catanionic nanovesicle, Hybrid lipid nanovesicle, Effective delivery, Bioavailability, Anticancer efficacy

1 Introduction

Lipid-based drug delivery systems enhance bioavailability of poorly water-soluble drugs by facilitating dissolution, enhancing the uptake and inhibiting the efflux transporters [1–3]. First-generation lipid carrier [4] were originally composed of pure solid lipids. Due to the intact crystal structure of solid lipid, the drug loading capacity of this type of nanoparticles is low, resulting in drug leakage during storage and limiting its usefulness as a drug delivery

Xuemei Xie and Dan He contributed equally to this work.

Kumaran Narayanan (ed.), *Bio-Carrier Vectors: Methods and Protocols*, Methods in Molecular Biology, vol. 2211,
https://doi.org/10.1007/978-1-0716-0943-9_5,

system. In order to overcome the shortcomings of solid lipid nanoparticles, hybrid lipid nanocarriers, also known as nanostructured lipid nanocarriers or the second-generation lipid nanocarriers, have been developed [5, 6].

The hybrid lipid nanocarriers are composed of a combination of solid lipid and liquid lipid. Since the liquid lipid is added to the nanocarrier, the lipids are present in a crystalline defect or amorphous structure state. As a result, hybrid lipid nanocarriers have a higher drug loading capability and suffer less drug leakage during preparation and storage compared to the pure lipid nanocarriers [7]. The catanionic nanovesicles are easily formed. The catanionic nanovesicles are thermodynamically stable systems composed of oppositely charged surfactants [8, 9].

In the research of antitumor delivery systems, the catanionic nanovesicles are attracting more and more attention, partly due to their ability to increase stability [10], adsorption [11], cellular uptake [12], apoptosis induction, tumor cell cytotoxicity, and antitumorigenic effect [13].

Curcumin (CC) is a yellow polyphenolic compound extracted from the rhizome of *Curcuma longa*. CC has been widely used to cure various cancers such as lung cancer, breast cancer, stomach cancer, and colon cancer [14]. However, CC has poor water solubility, easy photodegradation, chemical instability, low bioavailability, rapid metabolism, and fast systematic clearance. The unsatisfactory properties of CC severely hinder its clinical application [15].

In this chapter, we report the preparation of curcumin-loaded catanionic hybrid lipid nanovesicles (CC@CHLNV) (Fig. 1) evaluation of its pharmacokinetics, and anticancer properties in vitro and in vivo. CC@CHLNV consists of solid lipids, liquid lipids, and anionic surfactants and inherit the merits of hybrid lipid nanocarriers and catanionic nanovesicles. CC@CHLNV improved the solubility of CC, enhanced CC uptake, inhibited CC efflux, and greatly improved oral bioavailability of CC. Moreover, CC@CHLNV improved cell growth inhibition, apoptotic inducing and anti-invasion effects, and reduced cancerous growth.

2 Materials

Prepare all solutions using ultrapure water (25 °C, 18.2 Ω) and operate all processes at room temperature (unless indicated otherwise). Use analytical grade reagents unless indicated otherwise. Perform all animal experiments in accordance with the protocol approved by the Institutional Animal Ethics Committee.

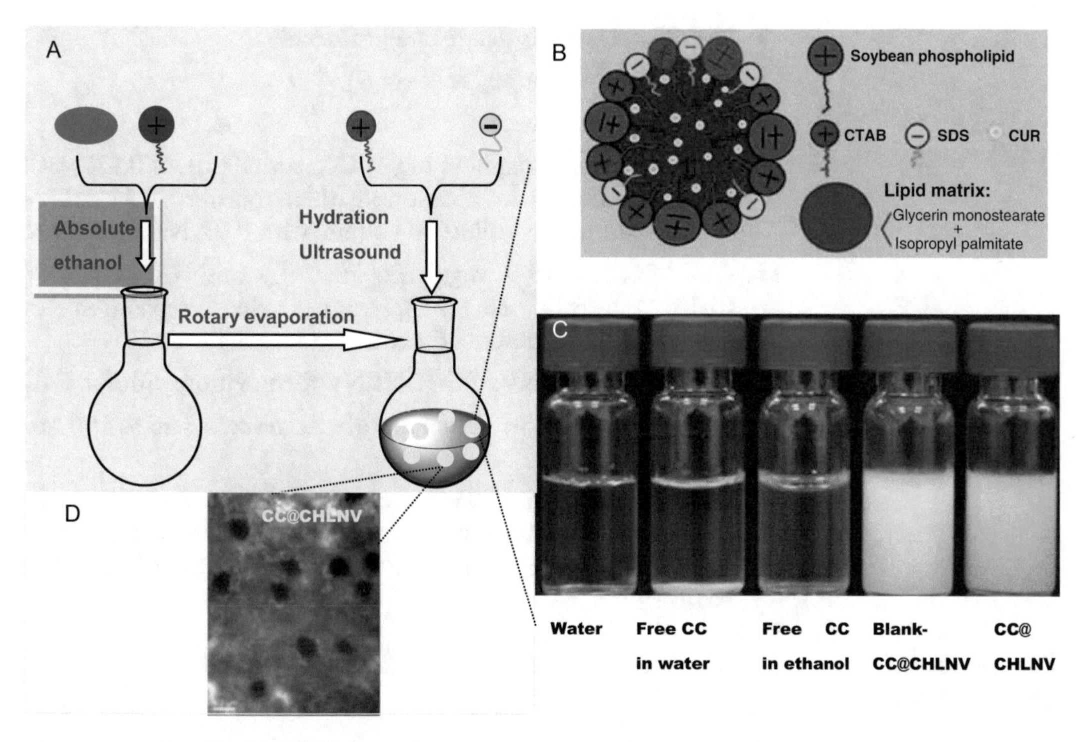

Fig. 1 Preparation and characteristics of CC@CHLNV. (**a**) Formulation prepared method: (1) add 10.00 mg CC, 70.24 mg glycerin monostearate, 54.07 mg isopropyl palmitate, 117.35 mg Lipoid S 7550 mL into round-bottomed flask with 20 mL of absolute ethanol and sonication for 2 min to obtain a yellowish solution (left flask). (2) evaporate solution at 30 °C under hypobaric conditions for 40 min to remove ethanol and obtain a CC@CHLNV yellow film (right flask). (**b**) schematic illustration of the CC@CHLNV structure. (**c**) optical photographs of water, free CC in water, free CC in ethanol, blank CC@CHLNV, and CC@CHLNV (from left to right). (**d**) transmission electron photomicrographs. CC@CHLNV tends to be circular and dispersed (bar: 200 nm)

2.1 Preparation of CC@CHLNV

1. Hybrid lipid matrix: Glycerin monostearate, Isopropyl palmitate (*see* **Note 1**).
2. Phospholipid: Lipoid S 75 (Phosphatidylcholine content: >70%, Phosphatidylcholine from soybean) (*see* **Note 2**).
3. Catanionic surfactants: Glycerin monostearate, cetyltrimethyl ammonium bromide (CTAB), and sodium dodecyl sulfate (SDS) (*see* **Note 3**).
4. 50 mL round-bottomed flask.

2.2 Pharmacokinetic Experiment

1. Sprague Dawley rats (230 ± 20 g).
2. Ether (≥99.5%).
3. Nitrendipine (>99.0%) (*see* **Note 4**).
4. Glacial acetic acid (≥99.5%) (*see* **Note 5**).
5. Acetonitrile (HPLC grade).
6. Ethyl acetate (≥99.5%).

7. 25 mL syringe, 16-gauge straight needle.
8. Brown sample bottle (*see* **Note 6**).
9. 0.9% NaCl solution.
10. CC suspension: Weigh 45 mg of CC, add 20 mL of 0.9% NaCl solution, dissolve by ultrasound, then transfer to 50 mL of brown volume, and dilute to volume with 0.9% NaCl solution.
11. CC@CHLNV (0.9 mg/mL, 50 mL) and CC@CHLNV (0.16 mg/mL, 15 mL), respectively. Preparation of CC@CHLNV is described in Subheading 3.1.
12. Blank CC@CHLNV: CC@CHLNV formulation without CC.
13. 5% glacial acetic acid: Add 5 mL of glacial acetic acid to 100 mL distilled water.
14. Mobile phase: Mix 550 mL acetonitrile and 450 mL 5% glacial acetic acid per 1000 mL. Pass the mobile phase through a 0.22 μm organic membrane and sonicate for 15 min (*see* **Note 7**).

2.3 Anticancer Efficiency In Vitro

1. Dulbecco's Modified Eagle's medium (DMEM).
2. 0.25% trypsin solution.
3. Dimethyl sulfoxide.
4. Penicillin-streptomycin double antibiotics.
5. PBS powder.
6. Fetal bovine serum.
7. MTT solution.
8. Annexin V-FITC-PI Apoptosis Detection Kit.
9. Propidium iodide (PI).
10. RNase A.
11. Calcium ion detection probe Fluo-3.
12. Reactive Oxygen Detection Probe DCFH.
13. Mitochondrial membrane potential detection probe.
14. Rhodamine123 (Rho123).
15. 96-well plates.
16. 6-well plates.
17. 15 mL centrifuge tube.
18. 1.5 mL eppendorf tube.
19. Complete medium: Add 50 mL fetal bovine serum and 5 mL penicillin-streptomycin double antibiotics into 445 mL DMEM medium.
20. PBS solution: Dissolve a bag of PBS powder in a 500 mL beaker and dilute to 1000 mL (*see* **Note 7**).

21. 70% cold ethanol: Mix 70 mL of absolute ethanol and 30 mL of distilled water, pre-cool at −20 °C.
22. Lewis lung cancer cells.

2.4 Anticancer Efficiency In Vivo

1. 1 mL syringe.
2. Electronic scale (ten thousandth balance).
3. Surgical Instruments (surgical scissors, tweezers ophthalmology).
4. C57BL/6J mice (20 ± 2 g).

3 Methods

3.1 Preparation Process of CC@CHLNV

1. Operate all steps in darkness (*see* **Note 6**). Weigh 10.00 mg CC, 70.24 mg glycerin monostearate, 54.07 mg isopropyl palmitate, 117.35 mg Lipoid S 75 (*see* **Note 8**). Transfer them to the 50 mL round-bottomed flask.
2. Add 20 mL of absolute ethanol to the same round-bottomed flask and sonicate for 2 min to obtain a yellowish solution.
3. Evaporate solution at 30 °C under hypobaric conditions for 40 min to remove ethanol and obtain a yellow film (*see* **Note 9**).
4. Take out the yellow powder and store them in 4 °C.
5. Add 241.66 mg of the yellow powder to 10 mL of distilled water containing CTAB and SDS (56.26 μmol:112.52 μmol) and sonicate for 10 min to obtain CC@CHLNV.

3.2 Pharmacokinetic Experiment

1. Randomly divide rats into four groups, fast for 12 h before administration (free drinking water), and orally administer with 0.9% NaCl (negative group), blank CC@CHLNV (Blank control group), 0.9 mg/mL CC suspension (treatment group 1), and CC@CHLNV (15 mg/kg, treatment group 2), respectively. Using a 5 mL syringe with 16-gauge straight needle, intravenously inject with 0.16 mg/mL CC suspension and CC@CHLNV (2.5 mg/kg) for the latter two group, respectively.
2. Analyze plasma CC concentration by HPLC. Under ether anesthesia, collect ophthalmic blood samples at 0, 5, 10, 15, 30, 45, 60, 120, 180 min, centrifuge at 6000 rpm for 10 min. Obtain plasma samples from the upper layers of supernatants, fix with 100 μL nitrendipine (5 μg/mL) and 1 mL of ethyl acetate. Vortex the mixture for 3 min and then centrifuge at 12,000 rpm for 10 min. Remove the organic layer and evaporate under nitrogen flow. Reconstitute the residue with 100 μL of the acetonitrile and 5% glacial acetic acid.

3. HPLC conditions include:

   Liquid chromatography column: Hypersil ODS-2 C18 column, 250 mm × 4:6 mm; 5 μm.

   Ultraviolet detector: 430 nm.

   Column temperature: 25 °C.

   Flow rate: 1 mL/min.

   Mobile phase: acetonitrile and 5% glacial acetic acid (55:45, v/v) (*see* **Note 10**).

3.3 Anticancer Efficiency In Vitro

3.3.1 Determination of Proliferation Inhibition on Lewis Lung Cancer Cells by MTT

1. Under the microscope, select the logarithmic growth phase cells, wash twice with sterile PBS solution, digest with 2 mL of 0.25% trypsin for 7 min, centrifuge at 1000 × *g* for 5 min, collect the cells, and add 4 mL of complete medium to prepare a cell suspension.
2. Cultivating Lewis lung cancer cells at a concentration of 3–5 × 10^3 cells in each well and inoculate them into 96-well plates.
3. Culture the inoculated cells in an incubator at 5% CO_2 and 37 °C for 12 h overnight until they were attached.
4. Set a negative control group, a free CC group, a CC@CHLNV group, and a blank CC@CHLNV group. Set the final drug concentrations at 2.5, 5.0, 10.0, 20.0, and 40.0 μmol/L. Set up six duplicate wells for each experimental group. Add 100 μL of every formulation diluted in complete medium to each well and continue culturing for 24 h.
5. Add 20 μL of MTT (5 mg/mL) to each well and continue to culture for 4 h.
6. Aspirate the supernatant and discard it, then add 150 μL dimethyl sulfoxide to each well. Shake the 96-well plate for 10 min on a shaker in the dark (*see* **Note 11**).
7. Set the enzyme-linked immunosorbent detector to a wavelength of 490 nm, and then measure the absorbance (A) of each well of a 96-well plate in the dark, three replicates each time (*see* **Note 12**). Cell inhibition rate (%) = $[1 - (A_{\text{Drug group}})/(A_{\text{Negative control group}})] \times 100\%$.

3.3.2 Determination of the Cell Cycle of Lewis Lung Cancer by Flow Cytometry

1. Under the microscope, select the logarithmic growth phase cells, wash twice with sterile PBS solution, digest with 2 mL of 0.25% trypsin for 7 min, centrifuge at 1000 × *g* for 5 min, collect the cells, and add 4 mL of complete medium to prepare a cell suspension.
2. Cultivate Lewis lung cancer cells at a concentration of 3–5 × 10^5 cells in each well and inoculate them into 6-well plates.

3. Culture the inoculated cells in an incubator at 5% CO_2 and 37 °C for 12 h overnight until they were attached.
4. Treat cells with 20 μmol/L Cc, CC@CHLNV, and blank CC@CHLNV.
5. Treat the cells for 24 h in an incubator at 5% CO_2 and 37 °C, digest the cells with 1 mL of 0.25% trypsin for 7 min, and transfer the cell suspension to a centrifuge tube and centrifuge at 1000 × *g* for 3 min.
6. Discard the supernatant, wash the cells with 2–3 times PBS solution, discard the supernatant, and add 70% cold ethanol to fix the cells at 4 °C for 24 h.
7. Add propidium iodide (PI) and RNase to the fixed cells, measure the cell cycle on a flow cytometer at 488 nm after incubation at 37 ° C for 30 min (*see* **Note 13**).

3.3.3 Determination of Cell Apoptosis of Lewis Lung Cancer by Flow Cytometry

1. Under the microscope, select the logarithmic growth phase cells, wash twice with sterile PBS solution, digest with 2 mL of 0.25% trypsin for 7 min, centrifuge at 1000 × *g* for 5 min, collect the cells, and add 4 mL of complete medium to prepare a cell suspension.
2. Cultivate Lewis lung cancer cells at a concentration of 3–5 × 10^5 cells in each well and inoculate them into 6-well plates.
3. Culture the inoculated cells in an incubator at 5% CO_2 and 37 °C for 12 h overnight until they are attached.
4. Treat cells with 20 μmol/L Cc, CC@CHLNV, and blank CC@CHLNV.
5. Incubate the cells for 24 h in an incubator at 5% CO_2 and 37 °C, digest the cells with 1 mL of 0.25% trypsin for 7 min, and transfer the cell suspension to a centrifuge tube and centrifuge at 1000 × *g* for 3 min.
6. Transfer the cell pellet to a centrifuge tube and add 195 μL Annexin V-FITC binding solution to prepare a suspension (*see* **Note 14**).
7. Add 5 μL of Annexin V-FITC staining solution to the EP tube, incubate on ice for 10 min in the dark.
8. Centrifuge at 1000 × *g* for 3 min, discard the supernatant, and add Annexin V-FITC binding solution to prepare a cell suspension.
9. Add 10 μL of PI staining solution, mix well into a cell suspension, incubate on ice for 10 min in the dark.
10. Detect the samples by flow cytometry (*see* **Note 15**).

3.3.4 Determination of Intracellular Active Oxygen Concentration by Flow Cytometry

1. Under the microscope, select the logarithmic growth phase cells, wash twice with sterile PBS solution, digest with 2 mL of 0.25% trypsin for 7 min, centrifuge at 1000 × *g* for 5 min, collect the cells, and add 4 mL of complete medium to prepare a cell suspension.
2. Cultivating Lewis lung cancer cells at a concentration of 3–5 × 10^5 cells in each well and inoculate them into 6-well plates.
3. Culture the inoculated cells in an incubator at 5% CO_2 and 37 °C for 12 h overnight until they were attached.
4. Treat the cells with 20 μmol/L CC, CC@CHLNV, and blank CC@CHLNV and culture the cells in an incubator at 5% CO_2 and 37 °C for 24 h, remove the complete medium, add DCFH-DA dye (10 μmol/L) to the solution and incubate for 20 min in the incubator at 5% CO_2, 37 °C (*see* **Note 14**).
5. Wash the cells three times with medium containing no serum.
6. Measure the change in fluorescence intensity of the cell sample by flow cytometry. Set the excitation wavelength to 488 nm and the emission wavelength to 525 nm.

3.3.5 Determination of the Intracellular Calcium Ion Concentration by Flow Cytometry

1. Under the microscope, select the logarithmic growth phase cells, wash twice with sterile PBS solution, digest with 2 mL of 0.25% trypsin for 7 min, centrifuge at 1000 × *g* for 5 min, collect the cells, and add 4 mL of complete medium to prepare a cell suspension.
2. Cultivating Lewis lung cancer cells at a concentration of 3–5 × 10^5 cells in each well and inoculate them into 6-well plates.
3. Culture the inoculated cells in an incubator at 5% CO_2 and 37 °C for 12 h overnight until they were attached.

   The operation steps are the same as those in Subheading 3.3.2, **steps 1–3**.
4. Treat the cells with 20 μmol/L CC, CC@CHLNV, and blank CC@CHLNV and culture the cells in an incubator at 5% CO_2 and 37 °C for 24 h, remove the medium, digest the cells by adding 2 mL of 0.25% trypsin for 7 min to cell suspension.
5. Centrifuge at 1000 × *g* for 3 min, discard the supernatant, wash the cells three times with PBS.
6. Add Fluo-3 fluorescent probe to the cell suspension and allow staining for 30 min (*see* **Note 14**).
7. Measure the change in fluorescence intensity of the cell sample by flow cytometry. Set the excitation wavelength to 488 nm and the emission wavelength to 525 nm.

3.3.6 Determination of Mitochondrial Membrane Potential by Flow Cytometry

1. Under the microscope, select the logarithmic growth phase cells, wash twice with sterile PBS solution, digest with 2 mL of 0.25% trypsin for 7 min, centrifuge at 1000 × *g* for 5 min, collect the cells, and add 4 mL of complete medium to prepare a cell suspension.
2. Cultivating Lewis lung cancer cells at a concentration of 3–5 × 10^5 cells in each well and inoculate them into 6-well plates.
3. Culture the inoculated cells in an incubator at 5% CO_2 and 37 °C for 12 h overnight until they were attached.
4. Treat the cells with 20 μmol/L CC, CC@CHLNV, and blank CC@CHLNV and culture the cells in an incubator at 5% CO_2 and 37 °C for 24 h, remove the medium, digest the cells by adding 2 mL of 0.25% trypsin for 7 min to cell suspension.
5. Add Rhodamine123 fluorescent probe to the cell suspension, let stain for 30 min (*see* **Note 14**).
6. Measure the change in fluorescence intensity of the cell sample by flow cytometry. Set the excitation wavelength to 488 nm and the emission wavelength to 525 nm.
7. Stain the tumor cells with Rhodamine, Fluo 3, and DCFH for 0.5 h and determine the fluorescence intensity using flow cytometry FACS Vantage (*see* **Note 18**).

3.4 Antitumor Efficiency In Vivo

1. Randomly divide tumor-bearing mice into four groups. Intraperitoneally give 0.9% NaCl, blank CC@CHLNV, CC or CC@CHLNV using a 1 mL syringe to each group. Intraperitoneally administer with 50 μg CC once every 48 h for each animal. After continuous administration for 14 days, observe the body condition of the mice every day (*see* **Note 19**), record the weight, and measure the size of the transplanted tumor with a vernier caliper.
2. Cut the external skin along the root of the tumor with surgical scissors, peel off the tumor tissue slowly from the skin using ophthalmic tweezers and surgical scissors. Obtain tumor tissue and record its weight. Sacrifice the mice by cervical dislocation. Isolate the tumor tissue with a surgical scissor, weight tumor tissue with a ten thousandth balance. (*see* **Note 20**).

4 Notes

1. Glycerin monostearate is used as solid lipid, and isopropyl palmitate is used as liquid lipid substance.
2. Lipoid S 75 is used as zwitter-ionic surfactant and emulsifying agent.

3. Glycerin monostearate and CTAB are used as positive-charged surfactant, and SDS is used as negative-charged surfactant.
4. In the pharmacokinetic experiment, due to the low content of CC, the addition of the internal standard nitrendipine during the measurement can improve the accuracy and precision of the measurement results. The basis of nitrendipine as the standard of CC internal standard: Nitrendipine is similar to CC in physical and chemical properties; it does not react with CC sample; it has similar peak time with CC and can be completely separated from components in CC sample.
5. Adding glacial acetic acid to the mobile phase can improve the peak shape of CC.
6. Light-sensitive CC should be preserved in the dark with aluminum foil.
7. Impurities and bubbles are removed by 0.22 μm organic filtration and sonication, respectively.
8. Encapsulation ratio and drug content of CC@CHLNV are influenced by the ratios of amounts of glycerin monostearate, isopropyl palmitate, Lipoid-S 75, and surfactant. In addition, the size and zeta potential of the nanoparticles may be changed by above materials. The ratios of materials in "Methods" are the optimal conditions. The optimal prescription is optimized by central composite design-response surface methodology to achieve the maximum encapsulation rate and drug loading.
9. Rotating evaporation time is related to the temperature, amounts of absolute ethanol and lipid materials. In general, 20 mL absolute ethanol and 200 mg lipid materials at 30 °C under hypobaric conditions for 40 min.
10. Under the above conditions, the peak times of CC and nitrendipine are 5.8 min and 7.6 min, respectively, and the peak shape is good and there is no interference. Due to the complex composition of the blood samples, the relatively high content of impurities, and the low concentration of CC in the blood samples, and the column is washed with mobile phase for 1–2 h after 15–20 CC samples are measured (the specific rinsing time is related to the sample and the column). This behavior may improve the accuracy of CC content determination and increase the accuracy of calculating CC bioavailability.
11. Oscillating for 10 min allow the drug to dissolve better.
12. The error is small when the absorbance is between 0.2 and 0.8. If it is beyond this range, adjust the number of inoculated cells and incubation time.

13. CC, CC @ CHLNV, and blank CC @ CHLNV arrests Lewis lung cancer cell cycle in S phase. The proportions of cells in S phase after treatment with CC, CC @ CHLNV, and blank CC @ CHLNV groups should be (81.57 ± 9.15)%, (92.20 ± 7.65)%, and (64.43 ± 10.12)%, respectively.
14. There are quenching problems in fluorescent dyes. Keep in the dark during storage and use to slow down fluorescence quenching.
15. In the early stage of apoptosis, different types of cells will turn phosphatidylserine out to the cell surface. Annexin V selectively binds phosphatidylserine. Annexin V labeled with FITC (Annexin V-FITC), a green fluorescent probe, can be used to detect the eversion of phosphatidylserine, an important feature of apoptosis, directly and simply by flow cytometry. Propidium iodide can stain the cells that lose the integrity of cell membrane in the late stage of apoptosis, showing red fluorescence.
16. In the CC group and CC@CHLNV group, the intracellular reactive oxygen species fluorescence intensity of Lewis lung cancer cells should be (2088.01 ± 238.19) and (2568.67 ± 361.85), respectively.
17. In the CC group and CC@CHLNV group, the intracellular calcium ion fluorescence intensities of Lewis lung cancer cells are expected to be (52.11 ± 11.28) and (97.85 ± 15.68), respectively.
18. In the CC group and CC@CHLNV group, the expected mitochondrial membrane potential fluorescence intensities of Lewis lung cancer cells are (128.97 ± 18.56) and (94.00 ± 12.11), respectively.
19. Weigh and record the body weight of the mice every other day.
20. Expected results: The tumor volume of CC@CHLNV group should decrease with prolonged administration time, and the tumor mass of CC@CHLNV group should shrink compared to other groups.

References

1. Yan S, Liu Y, Feng J, Zhao H, Yu Z, Zhao J, Li Y, Zhang J (2018) Difference and alteration in pharmacokinetic and metabolic characteristics of low-solubility natural medicines. Drug Metab Rev 50:140–160
2. Wan K, Sun L, Hu X, Yan Z, Zhang Y, Zhang X, Zhang J (2016) Novel nanoemulsion based lipid nanosystems for favorable in vitro and in vivo characteristics of curcumin. Int J Pharm 504:80–88
3. Tan Q, Liu S, Chen X, Wu M, Wang H, Yin H, He D, Xiong H, Zhang J (2012) Design and evaluation of a novel evodiamine-phospholipid complex for improved oral bioavailability. AAPS PharmSciTech 13:534–547
4. Zhang JQ, Liu J, Li XL, Jasti BR (2007) Preparation and characterization of solid lipid nanoparticles containing silibinin. Drug Deliv 14:381–387

5. Carvajal-Vidal P, Fabrega MJ, Espina M, Calpena AC, Garcia ML (2019) Development of Halobetasol-loaded nanostructured lipid carrier for dermal administration: optimization, physicochemical and biopharmaceutical behavior, and therapeutic efficacy. Nanomedicine 20:102026
6. Sun L, Wan K, Hu X, Zhang Y, Yan Z, Feng J, Zhang J (2016) Functional nanoemulsion-hybrid lipid nanocarriers enhance the bioavailability and anti-cancer activity of lipophilic diferuloylmethane. Nanotechnology 27:085102
7. Liu B, He D, Wu J, Sun Q, Zhang M, Tan Q, Li Y, Zhang J (2017) Catan-ionic hybrid lipidic nano-carriers for enhanced bioavailability and anti-tumor efficacy of chemodrugs. Oncotarget 8:30922–30932
8. Srivastava A, Liu C, Lv J, Kumar Deb D, Qiao W (2018) Enhanced intercellular release of anticancer drug by using nano-sized catanionic vesicles of doxorubicin hydrochloride and gemini surfactants. J Mol Liq 259:398–410
9. Russo Krauss I, Imperatore R, De Santis A, Luchini A, Paduano L, D'Errico G (2017) Structure and dynamics of cetyltrimethylammonium chloride-sodium dodecylsulfate (CTAC-SDS) catanionic vesicles: high-value nano-vehicles from low-cost surfactants. J Colloid Interface Sci 501:112–122
10. Li S, Fang C, Zhang J, Liu B, Wei Z, Fan X, Sui Z, Tan Q (2016) Catanionic lipid nanosystems improve pharmacokinetics and anti-lung cancer activity of curcumin. Nanomedicine 12:1567–1579
11. Hu J, Sun L, Zhao D, Zhang L, Ye M, Tan Q, Fang C, Wang H, Zhang J (2014) Supermolecular evodiamine loaded water-in-oil nanoemulsions: enhanced physicochemical and biological characteristics. Eur J Pharm Biopharm 88:556–564
12. Liu S, Chen D, Yuan Y, Zhang X, Li Y, Yan S, Zhang J (2017) Efficient intracellular delivery makes cancer cells sensitive to nanoemulsive chemodrugs. Oncotarget 8:65042–65055
13. Zhao J, Liu S, Hu X, Zhang Y, Yan S, Zhao H, Zeng M, Li Y, Yang L, Zhang J (2018) Improved delivery of natural alkaloids into lung cancer through woody oil-based emulsive nanosystems. Drug Deliv 25:1426–1437
14. Fotticchia T, Vecchione R, Scognamiglio PL, Guarnieri D, Calcagno V, Di Natale C, Attanasio C, De Gregorio M, Di Cicco C, Quagliariello V, Maurea N, Barbieri A, Arra C, Raiola L, Iaffaioli RV, Netti PA (2017) Enhanced drug delivery into cell cytosol via glycoprotein H-derived peptide conjugated nanoemulsions. ACS Nano 11:9802–9813
15. Mehanny M, Hathout RM, Geneidi AS, Mansour S (2016) Exploring the use of nanocarrier systems to deliver the magical molecule; curcumin and its derivatives. J Control Release 225:1–30

Chapter 6

Magnetic Nanoparticles as a Carrier of dsRNA for Gene Therapy

Małgorzata Grabowska, Bartosz F. Grześkowiak, Katarzyna Rolle, and Radosław Mrówczyński

Abstract

Glioma belongs to the most aggressive and lethal types of cancer. Glioblastoma multiforme (GBM), the most common type of malignant gliomas, is characterized by a poor prognosis and remains practically incurable despite aggressive treatment such as surgery, radiotherapy, and chemotherapy. Brain tumor cells overexpress a number of proteins that play a crucial role in tumorigenesis and may be exploited as therapeutic targets. One such target can be an extracellular matrix glycoprotein—tenascin-C (TN-C). Downregulation of TN-C by RNA interference (RNAi) is a very promising strategy in cancer therapy. However, the successful delivery of naked double-stranded RNA (dsRNA) complementary to TN-C sequence (ATN-RNA) requires application of delivery vehicles that can efficiently overcome rapid degradation by nucleases and poor intracellular uptake. Here, we present a protocol for application of MNP@PEI as a carrier for ATN-RNA to GBM cells. The obtained complexes consisted of polyethyleneimine (PEI)-coated magnetic nanoparticles combined with the dsRNA show high efficiency in ATN-RNA delivery, resulting not only in significant TN-C expression level downregulation, but also impairing the tumor cells migration.

Key words Magnetic nanoparticles, RNAi, Tenascin-C, Glioblastoma, Gene therapy, Cytotoxicity, Transfection, Double-stranded RNA

1 Introduction

The most common type of glioma, accounting for 55% of brain tumors, is glioblastoma multiforme (GBM). It is rated as the most aggressive, a fourth grade cancer, according to the World Health Organization classification [1]. Recommended therapies include surgical resection, radiotherapy, and systemic chemotherapy with oral prodrug temozolomide. Depending on the therapy, a median survival rate amounts 12–15 months, while failure to attempt treatment results in patient's death within 3 months [2].

None of the current state-of-the-art treatments for malignant glioma can be regarded as effective. Complete surgical resection of

Kumaran Narayanan (ed.), *Bio-Carrier Vectors: Methods and Protocols*, Methods in Molecular Biology, vol. 2211, https://doi.org/10.1007/978-1-0716-0943-9_6,

GBM is often not feasible due to its location in significant areas of the brain. GBM, characterized by high degree of invasiveness, readily infiltrate surrounding tissues, leading to later disease progression or recurrence [3]. It is therefore important to limit the migration potential of GBM cells. To initiate the migration process glioma cells degrade the surrounding extracellular matrix (ECM) into a migration favorable microenvironment [4].

Tenascin-C (TN-C), one of the ECM proteins responsible for adhesion and invasiveness of cancer cells, is highly overexpressed in GBM compared to healthy tissues, what makes it a great therapeutic target [5]. The identification of new molecular biomarkers in GBM as well as new drug delivery strategies are now a major therapeutic challenge. One of the most promising cancer treatment methods is gene therapy based on RNA interference (RNAi), referring to post-transcriptional gene silencing mediated by either degradation or translation inhibition of target RNA.

RNAi is triggered by the introduction of double-stranded RNA (dsRNA) into a cell [6]. In experimental approach, dsRNA with sequence homologous to TN-C mRNA was used to reduce TN-C expression [7]. Use of this RNA agent, named ATN-RNA, administrated locally into the tumor's cavity during standard neurosurgical procedure, results in increased survival and better quality of life of patients [8]. To overcome the lack of an effective delivery method for dsRNA and the instability of the nucleic acids during the delivery, we successfully tested the magnetic nanoparticles modified with polyethyleneimine (MNP@PEI) as a therapeutics' carrier [9].

Here, we present a protocol for application of MNP@PEI as a carrier for ATN-RNA to GBM cells. We demonstrate the protocols for magnetic nanoparticles synthesis, for assessment of cytotoxicity of obtained materials, preparation of nanoparticles-dsRNA complexes and their application in delivery of dsRNA to GBM cells. We also describe a method for assessing the gene silencing level and an approach for detection of the cell migration impairment.

2 Materials

2.1 Magnetic Nanoparticles Synthesis

1. Working solution: 0.5 g of 25-kDa branched polyethyleneimine (PEI-25_{Br}), 250 μL of Capstone FS-65 (Du Pont) fluorosurfactant, 2.5 mL of NH_4OH, 10 mL of water.

2.2 Preparation of ATN-RNA

2.2.1 In Vitro Transcription of ATN-RNA

1. pT7/T3α18 plasmid harboring ATN-RNA sequence. ATN-RNA was chemically synthesized form eight short (approximately 40 nt) overlapping oligodeoxynucleotides: four homologous to target fragment, and four complementary to it. Complementary fragments overlap with a shift of 10 nt. Resulting 5′ overhangs contain sites recognized by restriction enzymes *Hind*III for sense strand and *Eco*RI for template strand.
2. T3 transcription kit (Thermo Fisher Scientific).
3. T7 transcription kit (Thermo Fisher Scientific).
4. Gel DNA purification kit.
5. DNA gel loading dye.
6. 10× TBE buffer: 1 M Tris base, 1 M boric acid, 0.02 M EDTA, in H_2O.
7. 1% agarose gel.
8. Horizontal gel electrophoresis system.
9. Transilluminator.

2.2.2 dsRNA Hybridization

1. Hybridization buffer: 20 mM Tris–HCl pH 7.5, 50 mM NaCl, in H_2O.
2. 10% PAGE with 7 M urea: 4 mL of 10× TBE buffer, 10 mL of 20% acrylamide/bisacrylamide, 19.2 g of urea, up to 40 mL of H_2O.
3. RNA gel loading dye.
4. SYBR Safe.
5. Transilluminator.

2.3 MNP@PEI/dsRNA Complex Preparation

1. ATN-RNA solution in water at a concentration of 100 ng/μL.
2. MNP@PEI solution in water at a concentration of 1000 μg Fe/mL.

2.3.1 Gel Retardation Assay

1. DNA gel loading dye.
2. 10× TBE buffer: 1 M Tris base, 1 M boric acid, 0.02 M EDTA.
3. 1% agarose gel.
4. Horizontal gel electrophoresis system.
5. System for gel documentation.
6. Multidimensional images software (e.g., Digital Imaging and Analysis System II, Serva).

2.3.2 UV–Vis Spectrophotometry

1. Microvolume spectrophotometer (e.g. NanoDrop 2000, Thermo Scientific).

2.4 Cell Culture Maintenance

1. U-118 MG cell line (ATCC).
2. Dulbecco's Modified Eagle Medium (DMEM) High Glucose.
3. Fetal bovine serum (FBS).
4. Penicillin-streptomycin.
5. Trypsin.
6. Phosphate-buffered saline (PBS).
7. T-75 flasks.
8. CO_2 incubator.

2.5 Visualization of Transfection Complexes in the Cells

1. Atto 550 NHS solution (10 mg/mL): 1 mg of Atto 550 NHS, 100 μL of DMSO. Store at −20 °C.
2. Borate buffer: 0.1 M borate, pH 8.5, 1 M NaOH in H_2O.
3. Slide-A-Lyzer G2 Dialysis Cassettes, 3.5 k MWCO, 3 mL (Thermo Fisher Scientific).
4. 8-well Nunc Lab-Tek Chamber Slide (Thermo Fisher Scientific).
5. Label IT Tracker Intracellular Nucleic Acid Localization Kit, Fluorescein (Mirus).
6. 4% formaldehyde solution: 1 mL of 16% formaldehyde, 3 mL of Dulbecco's phosphate-buffered saline containing $MgCl_2$ and $CaCl_2$ (DPBS).
7. Concanavalin A-FITC (Thermo Fischer) solution (1 mg/mL): 10 mg of Concanavalin A-FITC, 10 mL of 0.1 M sodium bicarbonate. Aliquot and store at −20 °C.

2.6 Cytotoxicity Assays

1. Black polystyrene 96-well flat bottom μClear plate with the transparent bottom (Greiner Bio-One GmbH).
2. Live/Dead assay solution: 2 μM calcein AM, 2 μM ethidium homodimer-1, 8 μM Hoechst 33342 in DPBS. Protect from light, storing in a tube wrapped with aluminum foil.

2.7 Transfection

1. DMEM High Glucose.
2. PBS.
3. 12-well cell culture plate.

2.8 Establishing TNC-C Expression Level

2.8.1 Total RNA Isolation from Cell Culture

1. TRIzol Reagent.
2. Chloroform.
3. 2-Propanol.
4. 75% ethanol.
5. Refrigerated microfuge.
6. Microvolume spectrophotometer.

2.8.2 Removal of Residual DNA and Reverse Transcription

1. DNA removal kit (Ambion).
2. cDNA reverse transcription kit (Roche).
3. RNA gel loading dye.
4. 10× TBE buffer: 1 M Tris base, 1 M boric acid, 0.02 M EDTA.
5. 1% agarose gel.
6. Vertical gel electrophoresis system.
7. Transilluminator.
8. Microvolume spectrophotometer.

2.8.3 Real-Time Polymerase Chain Reaction

1. RT-PCR probe mix 2×.
2. Primers for the gene of interest.
3. Primers for the endogenous gene.
4. Probe from the Universal Probe Library (UPL, Roche).
5. 96-well PCR plates.
6. Adhesive PCR plate foils.
7. Real-time PCR detection system.

2.9 Cell-Migration Analysis

1. DMEM High Glucose.
2. PBS.
3. 12-well cell culture plate.
4. 200 μL micropipette tip.
5. Microscope with ×5 magnification.
6. Software with edge-detection algorithm to wound-healing assays analysis (e.g., TScratch, CSElab).

3 Methods

3.1 Synthesis and Functionalization of Magnetic Nanoparticles

1. Mix $FeCl_3{\cdot}6H_2O$ (135 mg, 0.5 mmol) with $FeCl_2{\cdot}4H_2O$ (50 mg, 0.25 mmol) in 5 mL of water and degassed with N_2 (*see* **Note 1**).
2. Heat the mixture up to 80 °C. Add 1 mL of working solution and continue heating for 120 min. Cool down the mixture and collect nanoparticles by an external magnet (*see* **Note 2**).
3. Wash nanoparticles with water (2 × 150 mL) and collected by an external magnet. Finally, redisperse nanoparticles in 10 mL of water.

3.2 Preparation of ATN-RNA

3.2.1 In Vitro Transcription of ATN-RNA

1. Digest 50 μg of plasmid harboring ATN-RNA construct separately with restriction enzymes *Eco*RI and *Hind*III.
2. Carry out the digestion process by minimum 8 h or overnight.
3. Separate the whole reaction on the 1% agarose gel and examine it on the transilluminator.
4. From agarose gel cut out the digested form of plasmid and purify the plasmid from the gel using gel DNA purification kit.
5. Independently transcribe the two strands of RNA with T3 and T7 polymerases for 1 μg of EcoRI and HindIII digested plasmids, respectively. Incubate the reaction overnight.
6. Purify the reaction loading whole mixture onto spin column plugged into sterile 1.5 mL collection tubes, then centrifuge at 750 × *g*, 2 min.
7. Apply 20 μL of nuclease-free H_2O on spin tubes and repeat centrifugation to new collection tubes.
8. Store transcripts in −20 °C.

3.2.2 dsRNA Hybridization

1. Mix equal quantities of T3 and T7 transcripts. Add 20 μL of 5× hybridization buffer and fill up with H_2O to 100 μL.
2. Incubate: 95 °C, 3 min, then follow the incubation for 75 °C, 30 min.
3. Slowly cool down the reaction for minimum 4 h or overnight to 25 °C.
4. Check the quality of dsRNA by separating with accompanying T3 and T7 non-hybridized transcripts on 10% polyacrylamide denaturing gel.
5. Rinse the gel for half an hour in SYBR Safe in water solution to visualize the bands.

3.3 MNP@PEI/dsRNA Complexes Preparation

1. Thaw dsRNA on ice and stir it gently by pipetting.
2. Vortex vigorously nanoparticles to break up the sediment and scatter nanoparticles in ultrasonic disperser for 15 min.
3. Mix dsRNA with MNP@PEI in a series of Fe to RNA weight ratios (1, 2, 3, 4, 5, 8, 10 Mag@PEI to ATN-RNA wt:wt ratios). Manipulate only with the amount of nanoparticles, as each complex must have an identical amount of dsRNA.
4. Adjust the volume of the complexes up to 50 μL with nuclease-free water.
5. Incubate: room temperature, 30 min.
6. Determine the rate of binding by gel retardation assay (Subheading 3.3.1) or UV-VIS spectrophotometry (Subheading 3.3.2). Use the free RNA as a control.

3.3.1 Gel Retardation Assay

1. Mix 15 μL of the complexes with 3 μL of gel loading dye solution and load MNP@PEI/dsRNA complexes on 1% agarose gel. As a control use the pure dsRNA.
2. Run electrophoresis in 1× TBE buffer at a constant voltage 10 V/cm for 30 min.
3. Visualize the samples using gel imaging system.
4. Calculate the binding efficiency by comparing density data obtained for complexes to those for the pure dsRNA.

3.3.2 UV-Vis Spectrophotometry

1. Centrifuge MNP@PEI/dsRNA complexes: 9300 × *g*, 10 min.
2. Measure the concentration of RNA contained in the supernatant at A260 on the spectrophotometer. Load 1 μL of the supernatant on the measurement pedestal and measure the absorbance.
3. Measure the concentration of pure dsRNA sample.
4. Calculate binding efficiency by comparing concentrations obtained for complexes to those for the pure dsRNA.

3.4 Cell Culture Maintenance

1. Grow adherent U-118 MG glioblastoma multiforme cell line in T-75 flask. Use culture medium supplemented with 10% FBS and 1% antibiotic solution, then replace it at intervals 24–36 h.
2. To passage the culture detach cells using trypsin. Carry out the culture in CO_2 incubators at temperature 37 °C, humidity 95%, CO_2 concentration 5% (*see* **Note 3**).

3.5 Visualization of Transfection Complexes in the Cells

3.5.1 Cellular Uptake of Transfection Complexes

1. To obtain fluorescently labeled MNP@PEI, mix 2 mL of magnetic NPs suspension (2 mg Fe/mL) with 490 μL 0.1 M borate buffer and 10 μL solution of Atto 550 NHS.
2. Incubate the suspension overnight at room temperature with continuous stirring.
3. Dialyze the suspension against water using a Slide-A-Lyzer G2 cassette with a cut-off at 3500 MW. Determine the iron concentration in obtained sample on a spectrophotometer.
4. Seed U118-MG cells into 8-well Nunc Lab-Tek Chamber Slide at 2.5 × 10^4 cells/well (300 μL) and incubate under standard tissue culture conditions for 24 h.
5. Prepare working dilution of transfection complexes containing fluorescently labeled NPs with an iron-to-RNA ratio of 3:1 (w/w). Transfer 50 μL of the transfection complexes corresponding to the ATN-RNA final concentration of 100 nM into the well of slide with the seeded cells. Incubate the cells with complexes for 24 h.

6. Wash the cells 1× with DPBS and then fix the cells by adding 200 μL of 4% formaldehyde and incubate for 15 min at room temperature.
7. Wash the cells 1× with DPBS and label the cellular membranes by adding 200 μL of Concanavalin A-FITC at the concentration of 25 μg/mL and incubate for 30 min at room temperature. Protect from light.
8. Wash the cells 1× with DPBS and label the nuclei by adding 200 μL of Hoechst 33342 at the concentration of 8 μM and incubate for 10 min at room temperature. Protect from light.
9. Wash the cells 1× with DPBS. Samples may be stored in 300 μL at 4 °C.
10. Visualize the cells using confocal laser scanning microscope. For the experiments detailed in this section, the images were acquired on the Olympus FV1000 with a ×60 objective, a 1.4 oil immersion lens, and analyzed using the FV10-ASW software (*see* **Note 4**).

3.5.2 Colocalization of Magnetic Nanoparticles and RNA Inside the Cells

1. To obtain fluorescently labeled ATN-RNA with Label IT Tracker Intracellular Nucleic Acid Localization Kit, mix 5 μg of RNA with 10× Labeling Buffer A, *Label* IT® Reagent and DNase-, RNase-free water in a volume of 50 μL.
2. Quantify the final concentration on a spectrophotometer.
3. Seed U-118 MG cells into 8-well Nunc Lab-Tek Chamber Slide at 2.5 × 10^4 cells/well (300 μL) and incubate under standard tissue culture conditions for 24 h.
4. Prepare working dilution of transfection complexes containing Atto 550 NHS-labeled NPs (red) and fluorescein-labeled RNA (green) with an iron-to-RNA ratio of 3:1 (w/w). Transfer 50 μL of the transfection complexes corresponding to the ATN-RNA final concentration of 100 nM into the well of slide with the seeded cells. Incubate the cells with complexes for 24 h.
5. Wash the cells 1× with DPBS and then fix the cells by adding 200 μL of 4% formaldehyde and incubate for 15 min at room temperature. Protect from light.
6. Remove fixative from the cells and then wash the cells 1× with DPBS.
7. Label the nuclei by adding 200 μL of Hoechst 33342 at the concentration of 8 μM and incubate for 10 min at room temperature. Protect from light. Samples may be stored in 300 μL at 4 °C.
8. Visualize the cells using confocal laser scanning microscope. For the experiments detailed in this section, the images were

acquired on the Olympus FV1000 with a ×60 objective, a 1.4 oil immersion lens, and analyzed using the FV10-ASW software (*see* **Note 5**).

3.6 Cytotoxicity Assays

3.6.1 WST-1 Cell Proliferation Assay

1. Seed U-118 MG cells into 96-well plate at 1×10^4 cells/well (150 μL) and incubate under standard tissue culture conditions for 24 h.
2. Prepare dilutions of transfection complexes with an iron-to-RNA ratio of 3:1 (w/w) corresponding to the ATN-RNA final concentration of 100, 50, 25, 12.5, and 6.25 nM. Transfer 50 μL each of the transfection complexes into the culture plates with the seeded cells. Incubate the cells with complexes for 24 h. Test each dilution of transfection complex in triplicate.
3. Add 10 μL WST-1 Cell Proliferation Reagent to each well and incubate at 37 °C for 4 h. Use cell culture medium as a blank.
4. Transfer 100 μL medium to the fresh wells on 96-well plate and record the absorbance at 450 nm (reference wavelength 620 nm) using a multiwell plate reader. Use untransfected cells as a reference.
5. Express the cell viability as: Cell viability (%) = $(Abs_{sample} - Abs_{blank})/(Abs_{ref} - Abs_{blank}) * 100$. Abs_{sample}, Abs_{blank}, and Abs_{ref} are absorbances at the maximum of the WST-1 absorption spectrum registered for a sample, blank, and reference sample, respectively.

3.6.2 Live/Dead Cell Viability Assay

1. Seed U-118 MG cells into 96-well μClear plate at 1×10^4 cells/well (150 μL) and incubate under standard tissue culture conditions for 24 h.
2. Prepare dilutions of transfection complexes with an iron-to-RNA ratio of 3:1 (w/w) corresponding to the ATN-RNA final concentration of 100, 50, 25, 12.5, and 6.25 nM. Transfer 50 μL each of the transfection complexes into the culture plates with the seeded cells. Incubate the cells with complexes for 24 h. Test each dilution of transfection complex in triplicate.
3. Remove medium from all wells and replace with 100 μL of Live/Dead assay solution. Incubate at 37 °C for 30 min.
4. Scan the plate using a high-content imaging system. For the experiments detailed in this section, the images were acquired on the IN Cell Analyzer 2000 with a ×20 objective and analyzed using the IN Cell Developer Toolbox software using in-house developed protocol (*see* **Note 6**).

3.7 Transfection

1. Calculate the volumes of nanoparticles and dsRNA needed for transfection based on binding efficiency obtained from Subheadings 3.3.1 and 3.3.2. Determine what volume of dsRNA

should be delivered to the cells to obtain the desired concentration of RNA and the amount of nanoparticles needed to bind them.

2. Seed U-118 MG cells into 12-well plate at 1.5×10^5 cells/well (1 mL) and incubate under standard tissue culture conditions for 24 h to obtain 75–90% confluence.
3. Remove medium from the cells, then wash the cells with PBS and add the supplemented medium in volume reduced by the amount of the transfection mixture.
4. Prepare MNP@PEI and dsRNA and form the MNP@PEI/dsRNA complexes as described in Subheading 3.3.
5. Add prepared transfection complexes to the cells, slowly spotting them on the surface of the whole well and incubate the cells for 24 h.
6. Depending on the purpose of the experiment, go to Subheading 3.8 or 3.9.

3.8 Establishing TNC-C Expression Level

3.8.1 Total RNA Isolation from Cell Culture

1. Remove medium from the cells, then wash the cells with PBS.
2. Add 250 μL of TRIzol reagent per one well from 12-well plate.
3. Incubate plate: 10 min, room temperature.
4. Transfer the lysate to a sterile 1.5 mL tube. Add 50 μL of chloroform and mix by shaking for 15 s to extract RNA.
5. Incubate: 10 min, room temperature and centrifuge at 12,000 × *g*, 4 °C, 15 min.
6. Transfer the upper, aqueous phase containing the RNA into a new sterile 1.5 mL tube.
7. Add 125 μL of 2-propanol and mix by inverting the tube several times.
8. Incubate for 10 min at room temperature and centrifuge at 12,000 × *g*, 4 °C, 8 min.
9. Remove the supernatant and add 0.5 mL of cold 75% ethanol. Mix by vortexing and centrifuge at 7500 × *g*, 4 °C, 10 min.
10. Remove the supernatant. Leave the tube open for 5–10 min to evaporate the ethanol. Do not let the sediment to dry out completely.
11. Dissolve RNA in 20 μL of nuclease-free water. Keep the RNA on ice (*see* **Note 7**).
12. Measure the amount of obtained RNA at A260 on the spectrophotometer.

3.8.2 Removal of Residual DNA and Reverse Transcription

1. Dispense 5 μg of isolated RNA.
2. Remove the residual DNA from sample using DNA removal kit.

3. Measure the amount of purified RNA at A260 on the spectrophotometer.
4. Investigate the quality of RNA by electrophoretic separation in the 1% agarose gel. If the resulting material does not contain any degradation or DNA residue, go to **step 5**. If not then repeat DNA removal or reisolate the RNA.
5. Using the cDNA reverse transcription kit and the isolated, pure total RNA, perform the reverse transcription according to the manufacturer's instructions.
6. Use the final product as a template in Subheading 3.8.3.

3.8.3 Real-Time Polymerase Chain Reaction

1. Design primers and probes for endogenous gene and gene of interest by the Universal Probe Library Assay Design Center.
2. Dilute primers to the concentration 10 μM.
3. Estimate the efficiency of primers on the standard curve with series of two-fold dilutions. Only primers not differing by more than 10% of efficiency can be subjected to one analysis.
4. Prepare the reaction mix: 5 μL of RT-PCR probe mix 2×, 1 μL of forward primer, 1 μL of reverse primer, 0.1 μL of probe, 1.9 μL of sterile water. Make three technical replicates for each experiment.
5. Pipette the reaction mixture onto 96-well plate and add 1 μL of template cDNA to the proper wells.
6. Seal the plate with adhesive PCR plate foil, centrifuge at 1800 × *g*, 3 min and place the plate in a real-time PCR detection system.
7. Proceed reaction under the following conditions: 5 min of preincubation at 95 °C, 40 cycles of denaturation in 95 °C for 10 s, annealing at proper to primers temperature for 30 s and extension at 72 °C for 10 s.
8. Calculate the level of gene expression in the test samples by first comparing the Ct values obtained for the test gene to those obtained for the endogenous gene, and then comparing the test samples to the control.

3.9 Cell-Migration Analysis

1. Remove the transfection mix from the cells, then wash the cells with PBS.
2. Add 1 mL of supplemented medium.
3. Create a "wound" by scratching the straight line in monolayer of cells by 200 μL tip.
4. Using the ×5 magnification take an image of scratch area at 12 h intervals. The image needs to be taken at exactly the same place of the well.
5. Observe for 48 h or until the scars are healed.

6. Analyze the degree of scarring of individual "wounds" at subsequent time points using wound-healing assays analysis software.

4 Notes

1. Bubbling N_2 at for 20 min in order to prevent unwanted oxidation of nanoparticles during the synthesis. Remember to keep the mixture all the time under N_2 atmosphere during the synthesis.
2. The collection of nanoparticles by external magnet can be performed in the beaker rather than in round bottom flask. The flat bottom of beaker gives bigger surface to collect nanoparticles and allows to handle them during the washing steps.
3. All activities related to cell culture perform under sterile conditions underneath a laminar flow cabinet.
4. Images of the Mag@PEI NPs were visualized using 559 nm excitation and 570–590 nm emission filters. To visualize the cell membranes, 488 nm excitation and 495–545 nm emission filters were applied. The Hoechst fluorescence was detected using 405 nm excitation source and 425–475 nm emission filters.
5. Images of the Mag@PEI NPs were visualized using 559 nm excitation and 570–590 nm emission filters, whereas ATN-RNA was visualized using 488 nm excitation and 495–545 nm emission filters. The Hoechst fluorescence was detected using 405 nm excitation source and 425–475 nm emission filters.
6. Perform all operations involving RNA manipulations on ice, using nuclease-free water.
7. Using a real-time PCR system based on probes allows for greater specificity of detection than a system based on fluorescent dyes.

Acknowledgments

The financial support under project number 2016/21/B/ST8/00477 granted by The National Science Centre, Poland to S.Jurga is kindly acknowledged.

References

1. Dunn GP, Rinne ML, Wykosky J et al (2012) Emerging insights into the molecular and cellular basis of glioblastoma. Genes Dev 26:756–784
2. Stupp R, Mason WP, van den Bent MJ et al (2005) Radiotherapy plus concomitant and adjuvant temozolomide for glioblastoma. N Engl J Med 352:987–996
3. Davis ME (2016) Glioblastoma: overview of disease and treatment. Clin J Oncol Nurs 20: S2–S8
4. Lefranc F, Brotchi J, Kiss R (2005) Possible future issues in the treatment of glioblastomas: special emphasis on cell migration and the resistance of migrating glioblastoma cells to apoptosis. J Clin Oncol 23:2411–2422
5. Brösicke N, Faissner A (2015) Role of tenascins in the ECM of gliomas. Cell Adhes Migr 9:131–140
6. Mathupala SP, Guthikonda M, Sloan AE (2006) RNAi based approaches to the treatment of malignant glioma. Technol Cancer Res Treat 5:261–269
7. Zukiel R, Nowak S, Wyszko E et al (2006) Suppression of human brain tumor with interference RNA specific for tenascin-C. Cancer Biol Ther 5:1002–1007
8. Wyszko E, Rolle K, Nowak S et al (2008) A multivariate analysis of patients with brain tumors treated with ATN-RNA. Acta Pol Pharm 65:677–684
9. Grabowska M, Grześkowiak BF, Szutkowski K et al (2019) Nano-mediated delivery of double-stranded RNA for gene therapy of glioblastoma multiforme. PLoS One 14:e0213852

Part III

Peptide Vehicles

Chapter 7

Generation of Functional Insulin-Producing Cells from Mouse Embryonic Stem Cells Through Protein Transduction of Transcription Factors

Taku Kaitsuka and Kazuhito Tomizawa

Abstract

In this chapter, we describe a simple and unique method for the differentiation of mouse embryonic stem cells into insulin-producing cells. In addition to cytokines and growth factors, key transcription factors for pancreatic development are applied in this method through protein transduction technology. Furthermore, a combination of nanofiber plates and laminin coatings improves the yield of differentiated cells. The insulin-producing cells derived through this method express marker genes of mature β-cells and have an ability to secrete insulin; therefore, these cells are useful for fundamental studies on pancreatic development, drug development, and regenerative medicine for diabetes.

Key words Mouse embryonic stem cells, Pancreatic differentiation, Protein transduction, Insulin-producing cells, Insulin secretion

1 Introduction

Pluripotent stem cells, such as embryonic stem (ES) and induced pluripotent stem (iPS) cells, are capable of differentiating into somatic cells in vitro. Among these cells, pancreatic β-cells differentiated from pluripotent stem cells are useful for studies on pancreatic development, drug development, and regenerative medicine for diabetes. To date, various protocols for pancreatic differentiation have been reported and utilized for basic and applied research [1–4]. However, some of these protocols consist of numerous complicated steps and require many hormones, growth factors, and cytokines to guide the differentiation of the cells into insulin-producing cells.

We have established a virus-free method for pancreatic differentiation from mouse embryonic stem cells that is also simpler than the previous protocols [5–7]. In stage 1 of our method, the cells are treated with Activin A and basic fibroblast growth factor (bFGF)

Kumaran Narayanan (ed.), *Bio-Carrier Vectors: Methods and Protocols*, Methods in Molecular Biology, vol. 2211,
https://doi.org/10.1007/978-1-0716-0943-9_7,

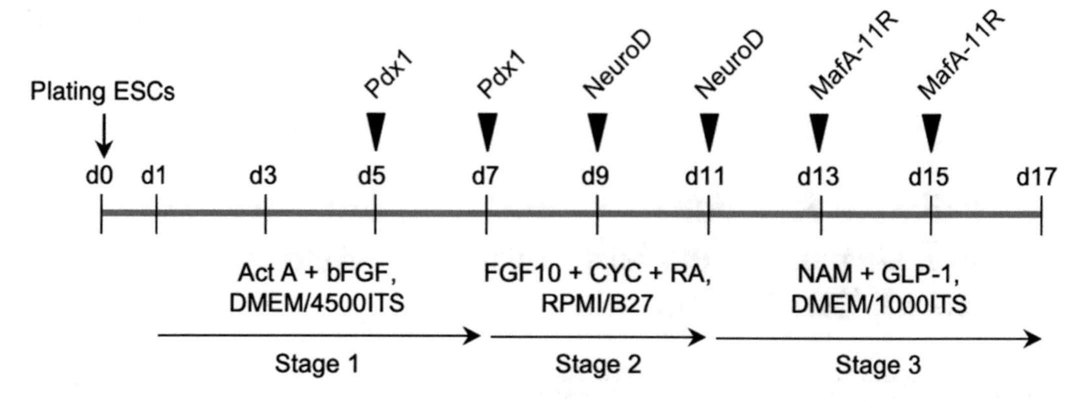

Fig. 1 A timeline of the differentiation protocol. Three types of differentiation medium are applied, as indicated. The recombinant proteins Pdx1, NeuroD, and MafA-11R should be added to the cells at the indicated time points. *11R* 11 arginines, *Act A* activin A, *bFGF* basic FGF, *CYC KAAD-cyclopamine, RA* retinoic acid, *NAM* nicotinamide, *GLP-1* glucagon-like peptide-1, *ITS* insulin-transferrin-selenium

and are then guided to become endodermal cells. In stage 2, the cells are treated with fibroblast growth factor 10 (FGF10), hedgehog inhibitor, and retinoic acid and are then guided to become pancreatic progenitors. In stage 3, the cells are treated with nicotinamide and glucagon-like peptide-1 (GLP-1) and are then guided to become insulin-producing cells. During these steps, three recombinant proteins, Pdx1, NeuroD, and MafA-11R, are treated at specific time points (Fig. 1).

Protein transduction technology is a method for the delivery of recombinant proteins or peptides to the inside of live cells [8–10]. Arginine- or lysine-rich sequences are defined as protein transduction domains (PTDs) and trigger the internalization of proteins into cells [11, 12]. While Pdx1 and NeuroD each have a PTD [13, 14], MafA does not; therefore, 11 arginines were fused to the C-terminus of this protein. While these three proteins are key transcription factors in pancreatic development, they emerge and function at different stages [15].

At the end of differentiation, the yield, maturity, and function of the differentiated cells can be assessed through immunocytochemistry, quantitative PCR analysis, and insulin secretion assays. After such quality checks, the cells can be used for further studies, such as studies on the effects of small molecules on insulin secretion and on the regulation of gene expression in pancreatic development.

2 Materials

Use Milli-Q-purified water or equivalent for the preparation of all buffers.

2.1 Purification of Recombinant Proteins

1. Plasmids: rat Pdx1-pET21a, rat NeuroD-pET21a, and mouse MafA-11R-pGEX-6p-1 (*see* **Note 1**).
2. 0.1 M Isopropyl-β-D-thiogalactopyranoside (IPTG): 0.1 M in water. Store at −20 °C.
3. Lysis buffer for histidine (His)-tag purification: 20 mM HEPES-NaOH (pH 8.0), 100 mM NaCl, 1% Triton X-100, and 0.1 mg/mL lysozyme. Store at 4 °C. Add 1 tablet/50 mL cOmplete EDTA-free protease inhibitor cocktail (Roche) before use.
4. Wash buffer for His-tag purification: 20 mM HEPES-NaOH (pH 8.0), 100 mM NaCl, and 20 mM imidazole. Store at 4 °C. Add 1 tablet/50 mL cOmplete EDTA-free protease inhibitor cocktail (Roche) before use.
5. Elution buffer for His-tag purification: 20 mM HEPES-NaOH (pH 8.0) 100 mM NaCl, and 500 mM imidazole. Store at 4 °C. Add 1 tablet/50 mL cOmplete EDTA-free protease inhibitor cocktail (Roche) before use.
6. Lysis buffer 1 for glutathione-S-transferase (GST)-tag purification: 50 mM Tris–HCl (pH 8.0), 300 mM NaCl, 1 mM EDTA, and 1 mM EGTA. Store at 4 °C.
7. Lysis buffer 2 for GST-tag purification: 50 mM Tris–HCl (pH 8.0), 300 mM NaCl, 1 mM EDTA, 1 mM EGTA, 4 mM DTT, 0.8% NP40, and 2 mg/mL lysozyme. Store at 4 °C. Add 1 tablet/25 mL cOmplete protease inhibitor cocktail (Roche) before use.
8. Wash buffer for GST-tag purification: 20 mM Tris–HCl (pH 8.0), 2 mM $MgCl_2$, and 200 mM NaCl. Store at 4 °C.
9. GST cleavage buffer: 50 mM Tris–HCl (pH 7.0), 150 mM NaCl, 1 mM EDTA, and 1 mM DTT. Store at 4 °C.

2.2 Culture of Embryonic Stem Cells

Make all media, buffers, and solutions sterile.

1. 0.1% Gelatin solution: Add 100 mL of water to a 100 mL medium bottle. Weigh 0.1 g of gelatin and transfer to the bottle. Mix and leave to stand for 30 min. Sterilize with an autoclave. Store at 4 °C.
2. Mouse embryonic fibroblast (MEF) medium: high-glucose (4500 mg/L) DMEM supplemented with 10% fetal bovine serum (FBS), 2 mM L-glutamine (L-Gln), 50 U/mL penicillin, and 50 μg/mL streptomycin. Store at 4 °C.
3. Embryonic stem cell (ESC) medium: Glasgow MEM supplemented with 1000 U/mL leukemia inhibitory factor (LIF), 15% knockout serum replacement (KSR; Thermo Fisher Scientific), 1% FBS, 100 μM nonessential amino acids (NEAAs),

2 mM L-Gln, 1 mM sodium pyruvate, 50 U/mL penicillin, 50 μg/mL streptomycin, and 100 μM β-mercaptoethanol (β-ME). Store at 4 °C.

2.3 Differentiation into Insulin-Producing Cells

Make all media, buffers, and solutions sterile.

1. Plate: 96-well plate with an Ultra-Web synthetic surface (Corning) or polyamide net-like fibers (Japan Vilene Company, Ltd.) (*see* **Note 2**).
2. Phosphate-buffered saline (PBS).
3. 1 mg/mL PLL stock solution: Dissolve 1 mg of poly-L-lysine in 1 mL of sterile PBS. Store at 4 °C.
4. 10 μg/mL PLL working solution: Mix 0.1 mL of PLL stock solution with 9.9 mL of sterile PBS. Store at 4 °C.
5. 10 μg/mL Laminin working solution: Mix 0.1 mL of laminin solution (laminin-111 from mouse Engelbreth-Holm-Swarm sarcoma; Roche) with 4.9 mL of sterile DMEM.
6. Plating medium: high-glucose (4500 mg/L) DMEM supplemented with 10% FBS, 100 μM NEAAs, 2 mM L-Gln, 50 U/mL penicillin, 50 μg/mL streptomycin, and 100 μM β-ME. Store at 4 °C.
7. 0.2 g/mL AlbuMAX II stock solution: Add approximately 100 mL of water to a glass beaker. Weigh 25 g of AlbuMAX II (Thermo Fisher Scientific) and transfer to the beaker. Mix and make up to 125 mL with water. Sterilize through a 0.22-μm filter. Store at −20 °C.
8. 100 μg/mL Activin A stock solution: Add 100 μL of sterile 0.1% bovine serum albumin (BSA)-PBS to a vial containing 10 μg of Activin A (R&D Systems) and mix. Store at −80 °C.
9. 50 μg/mL Basic fibroblast growth factor (bFGF) stock solution: Add 1 mL of sterile 0.1% BSA-PBS to a vial containing 50 μg of bFGF (PeproTech) and mix. Store at −80 °C.
10. 100 mM Retinoic acid (RA) stock solution: 100 mM RA in sterile dimethyl sulfoxide (DMSO). Store at −20 °C.
11. 250 μM Cyclopamine stock solution: 250 μM KAAD-cyclopamine (EMD Millipore) in sterile DMSO. Store at −20 °C.
12. 50 μg/mL Fibroblast growth factor 10 (FGF10) stock solution: Add 0.5 mL of sterile 0.1% BSA-PBS to a vial containing 25 μg of FGF10 (PeproTech) and mix. Store at −80 °C.
13. 1 M Nicotinamide (NAM) stock solution: 1 M in PBS. Sterilize through a 0.22-μm filter. Store at −20 °C.
14. 100 μM Glucagon-like peptide-1 (GLP-1) stock solution: 100 μM in sterile PBS. Store at −80 °C.

15. Stage 1 medium: high-glucose (4500 mg/L) DMEM supplemented with 1% ITS-G (Thermo Fisher Scientific), 2.5 mg/mL AlbuMAX II, 100 μM NEAAs, 2 mM L-Gln, 50 U/mL penicillin, 50 μg/mL streptomycin, 100 μM β-ME, 10 ng/mL Activin A, and 5 ng/mL bFGF. Store at 4 °C (*see* **Note 3**).
16. Stage 2 medium: RPMI supplemented with 100 μM NEAAs, 2 mM L-Gln, 50 U/mL penicillin, 50 μg/mL streptomycin, 100 μM β-ME, 2% B27 supplement (Thermo Fisher Scientific), 50 ng/mL FGF10, 250 nM KAAD-cyclopamine, and 1 μM RA. Store at 4 °C (*see* **Note 3**).
17. Stage 3 medium: low-glucose (1000 mg/L) DMEM supplemented with 1% ITS-G, 2.5 mg/mL AlbuMAX II, 100 μM NEAAs, 2 mM L-Gln, 50 U/mL penicillin, 50 μg/mL streptomycin, 100 μM β-ME, 10 mM NAM, and 10 nM GLP-1. Store at 4 °C (*see* **Note 3**).

2.4 C-peptide Secretion Assay

1. Krebs–Ringer solution containing bicarbonate and HEPES (KRBH) with low glucose: 129 mM NaCl, 4.8 mM KCl, 2.5 mM $CaCl_2$, 1.2 mM KH_2PO_4, 1.2 mM $MgSO_4$, 5 mM $NaHCO_3$, 10 mM HEPES-NaOH (pH 7.4), 0.1% BSA, and 2.5 mM D-glucose. Sterilize through a 0.22-μm filter. Store at 4 °C.
2. Krebs–Ringer solution containing bicarbonate and HEPES (KRBH) with high glucose: 129 mM NaCl, 4.8 mM KCl, 2.5 mM $CaCl_2$, 1.2 mM KH_2PO_4, 1.2 mM $MgSO_4$, 5 mM $NaHCO_3$, 10 mM HEPES-NaOH (pH 7.4), 0.1% BSA, and 20 mM D-glucose. Sterilize through a 0.22-μm filter. Store at 4 °C.

3 Methods

3.1 Purification of Recombinant Proteins (His-tag)

1. Culture the BL21 (DE3) bacteria carrying the desired plasmid (rat Pdx1-pET21a or rat NeuroD-pET21a) in 100 mL of LB medium with 100 μg/mL ampicillin overnight at 37 °C.
2. Decant the cultured bacteria into 900 mL of LB medium with 100 μg/mL ampicillin. Culture them again until the OD600 becomes approximately 0.8 at 37 °C (*see* **Note 4**).
3. Add 0.1 M IPTG at a final concentration of 0.2 mM to induce the expression of the desired protein. Continue the culture by incubating at 32 °C for 6–8 h (*see* **Note 5**).
4. Harvest the bacteria through centrifugation at 6000 × *g* for 10 min at 4 °C.
5. Add 100 mL of cold lysis buffer to the bacterial pellet and lyse the bacterial cells with a vortex mixer. Incubate the samples for 15 min on ice (*see* **Note 6**).

6. Sonicate the sample with an ultrasonicator (*see* **Note 7**). Centrifuge at 20,000 × *g* for 10 min at 4 °C and collect the supernatant in a new tube. Keep the sample on ice.
7. Wash 2 mL of TALON® resin (Clontech) with 10 mL of water. Centrifuge at 700 × *g* for 2 min at 4 °C and then remove the supernatant.
8. Equilibrate the resin with 10 mL of wash buffer. Centrifuge at 700 × *g* for 2 min at 4 °C and then remove the supernatant.
9. Apply the supernatant of the bacterial lysate to the equilibrated resin. Incubate the mixture with gentle rotation for 20 min at 4 °C. Centrifuge at 700 × *g* for 5 min at 4 °C and then remove the supernatant.
10. Wash the resin with 10 mL of cold wash buffer three times (*see* **Note 8**).
11. Elute the His-tagged protein with 12 mL of cold elution buffer.
12. Check the yield and purity of the eluted proteins by SDS-PAGE followed by Coomassie brilliant blue (CBB) staining.
13. Dialyze the protein by 4 L of PBS for 2 h at 4 °C and repeat twice (*see* **Note 9**). Concentrate the protein until the concentration reaches 20 μM by centrifugal filter units. Sterilize through a 0.22-μm filter. Store at −80 °C.

3.2 Purification of Recombinant Proteins (GST-Tag)

1. Culture the BL21 (DE3) bacteria carrying the desired plasmid (mouse MafA-11R-pGEX-6p-1) in 100 mL of LB medium with 100 μg/mL ampicillin overnight at 37 °C.
2. Decant the cultured bacteria into 900 mL of LB medium with 100 μg/mL ampicillin. Culture them again until the OD600 becomes approximately 0.8 at 37 °C (*see* **Note 4**).
3. Add IPTG at a final concentration of 0.2 mM to induce the expression of the desired protein. Continue the culture by incubating at 32 °C for 6–8 h (*see* **Note 5**).
4. Harvest the bacteria through centrifugation at 6000 × *g* for 10 min at 4 °C.
5. Add 50 mL of cold lysis buffer 1 to the bacterial pellets and resuspend them by pipetting and vortexing.
6. Add 50 mL of cold lysis buffer 2 to the bacteria and then lyse them with a vortex mixer. Incubate for 30 min on ice (*see* **Note 6**).
7. Sonicate the samples with an ultrasonicator (*see* **Note 7**). Centrifuge at 20,000 × *g* for 10 min at 4 °C and collect the supernatant in a new tube. Keep the sample on ice.

8. Wash 2.66 mL of Glutathione Sepharose 4 Fast Flow (GE) beads with 20 mL of water. Centrifuge at 500 × *g* for 2 min at 4 °C and then remove the supernatant.
9. Equilibrate the beads with 20 mL of wash buffer. Centrifuge at 500 × *g* for 2 min at 4 °C and then remove the supernatant.
10. Apply the supernatant of the bacterial lysate to the equilibrated beads and incubate them with gentle rotation for 1–2 h at 4 °C.
11. Wash the beads with 20 mL of cold wash buffer three times.
12. Transfer the beads to new 15 mL conical tubes. Equilibrate them with 10 mL of cold GST cleavage buffer. Centrifuge at 500 × *g* for 2 min at 4 °C and then remove the supernatant.
13. Prepare the PreScission protease (GE) mixture (*see* **Note 10**). Mix 60 μL of PreScission protease and 1940 μL of GST cleavage buffer.
14. Add 2 mL of the PreScission protease mixture to the beads and then rotate them for 4 h at 4 °C to cleave the GST-tag (*see* **Note 11**).
15. Centrifuge the beads at 500 × *g* for 5 min at 4 °C and then collect the supernatant into a new tube.
16. Elute the residual proteins with 6 mL of cold GST cleavage buffer. Centrifuge the beads at 500 × *g* for 5 min at 4 °C and then collect the supernatant.
17. Assess the yield and purity of the eluted proteins by SDS-PAGE followed by CBB staining.
18. Dialyze the protein by 4 L of PBS for 2 h at 4 °C and repeat twice (*see* **Note 9**). Concentrate the protein until the concentration reaches 20 μM by centrifugal filter units. Sterilize through a 0.22-μm filter. Store at −80 °C.

3.3 Preparation of Feeder Cells

1. Coat a culture dish with a 0.1% gelatin solution (2 mL/60 mm dish). Incubate for 30 min at 37 °C in a CO_2 incubator (*see* **Note 12**).
2. Thaw mitomycin C-treated MEFs (*see* **Note 13**) quickly at 37 °C in a water bath and transfer them to 15 mL conical tubes containing 5 mL of MEF medium. Centrifuge at 200 × *g* for 5 min and aspirate the medium. Resuspend the cells with 4 mL of MEF medium and plate them into gelatin-coated dishes at a density of 2~5 × 10^5 cells/60 mm dish.
3. Culture the cells for 24 h at 37 °C in a CO_2 incubator.

3.4 Culture of Embryonic Stem Cells

1. Thaw the ESCs (*see* **Note 14**) quickly at 37 °C in a water bath and transfer them to 15 mL conical tubes containing 5 mL of ESC medium. Centrifuge at 200 × *g* for 5 min and then

aspirate the medium. Resuspend the cells with 4 mL of ESC medium and plate them onto feeder cells in a 60 mm dish. Culture the cells at 37 °C in a CO_2 incubator.

2. Change the medium every day and continue culturing the cells until they become 80–90% confluent (*see* **Note 15**).

3.5 Coating of the Culture Plate with Poly-L-lysine and Laminin

1. Add 75 μL of the PLL working solution to the wells of a 96-well plate (0.75 μg/well) and incubate for more than 2 h at 37 °C in a CO_2 incubator (*see* **Note 16**).
2. Add 50 μL of the laminin working solution to the wells (0.5 μg/well) and incubate overnight at 37 °C in a CO_2 incubator (*see* **Note 17**).

3.6 Differentiation into Insulin-Producing Cells

A timeline of the differentiation is shown in Fig. 1.

1. Aspirate the medium of the ESCs and wash the cells with 2 mL of PBS. Add 1 mL of 0.25% trypsin-EDTA and incubate for 5 min at 37 °C in a CO_2 incubator.
2. Dissociate the cells with a P1000 micropipette to avoid aggregation of cells and then add 4 mL of plating medium.
3. Transfer the cells to 15 mL conical tubes and count them by a hemacytometer (*see* **Note 18**). Then, centrifuge at 200 × *g* for 5 min.
4. Add a defined amount of plating medium to the cell pellet to yield a suspension with 5.0×10^4 cells/mL. Resuspend the cells thoroughly by pipetting (*see* **Note 19**).
5. Before plating the cells, add 100 μL of plating medium to the PLL- and laminin-coated wells; then plate 100 μL of cells into each well to yield a density of 5000 cells/well (*see* **Note 20**). Culture the cells for 24 h at 37 °C in a CO_2 incubator.
6. One day after plating, change the medium to 200 μL of stage 1 medium (*see* **Note 21**).
7. Three days after plating, change the medium to 200 μL of stage 1 medium.
8. Five days after plating, change the medium to 200 μL of stage 1 medium. Add Pdx1 protein at a final concentration of 1 μM (*see* **Note 22**).
9. Seven days after plating, change the medium to 200 μL of stage 2 medium. Add Pdx1 protein at a final concentration of 1 μM.
10. Nine days after plating, change the medium to 200 μL of stage 2 medium. Add NeuroD protein at a final concentration of 1 μM.
11. Eleven days after plating, change the medium to 200 μL of stage 3 medium. Add NeuroD protein at a final concentration of 1 μM.

12. Thirteen days after plating, change the medium to 200 μL of stage 3 medium. Add MafA-11R protein at a final concentration of 1 μM.
13. Fifteen days after plating, change the medium to 200 μL of stage 3 medium. Add MafA-11R protein at a final concentration of 1 μM.

3.7 C-Peptide Secretion Assay

1. Wash the cells with 200 μL of low-glucose KRBH buffer once. Then, add another 200 μL of low-glucose KRBH buffer and incubate the cells for 1 h at 37 °C in a CO_2 incubator to wash out the residual medium (*see* **Note 23**).
2. Replace the buffer with 100 μL of fresh low-glucose KRBH buffer and incubate for 1 h at 37 °C in a CO_2 incubator. Transfer 80 μL of the buffer incubated with the cells to a 1.5 mL tube (*see* **Note 24**) and then aspirate the residual buffer. Keep the tube on ice.
3. Add 100 μL of high-glucose KRBH buffer and incubate for 1 h at 37 °C in a CO_2 incubator. Transfer 80 μL of the buffer incubated with the cells to a 1.5 mL tube (*see* **Note 25**) and then aspirate the residual buffer. Keep the tube on ice.
4. Add 20 μL of 1 N NaOH to the cells, lyse them by pipetting and then transfer them to a 1.5 mL tube. Centrifuge at 12,000 × *g* for 5 min at 4 °C and then transfer the supernatant to a new tube.
5. Measure the C-peptide levels in the buffers (low glucose and high glucose) incubated with the cells after centrifugation at 3000 × *g* for 5 min at 4 °C (*see* **Note 26**) using a mouse C-peptide ELISA kit (Shibayagi) (*see* **Note 27**).
6. Measure the total protein concentration of the supernatant of the lysed cells with Bradford reagent (*see* **Note 28**).
7. Divide the C-peptide levels by the total protein content to determine the secreted C-peptide as ng/protein (mg).

4 Notes

1. These plasmids are available upon request.
2. The 96-well plates with Ultra-Web synthetic surfaces have been discontinued. Plates with polyamide net-like fibers can be substituted for them [7]. For information about polyamide net-like fibers, please visit http://www.vilene.co.jp/.
3. The differentiation media supplemented with the factors should be used within 2 days.
4. The range as 0.6~0.8 of OD600 may work.

5. Alternatively, induce with 0.1 mM IPTG overnight at 16 °C.
6. In this step, it is not necessary to lyse the pellet thoroughly because the sonication procedure in the next step will disrupt it completely.
7. The sonication will produce the heat leading the degradation of proteins, therefore, this procedure should be done with cooling by ice.
8. Alternatively, the resin can be transferred to a disposable plastic column and then washed and eluted on the column via gravity-flow procedures.
9. Total time of dialysis should be between 6 and 8 h because long hours of it often cause a precipitation of proteins.
10. The backbone vector of mouse MafA-11R plasmid is pGEX-6p-1. It contains the recognition sequence for PreScission protease between the GST domain and the sequence encoding MafA-11R.
11. Overnight incubation can be performed if desired.
12. Gelatin solution should be aspirated before plating the feeder cells.
13. Treatment with mitomycin C should be performed at a concentration of 10 μg/mL for 2–3 h at 37 °C in a CO_2 incubator. Alternatively, those cells are commercially available.
14. We have tested the mouse ESC lines E14tg2a (provided by the RIKEN BRC) [16], Sk7 (carrying a Pdx1-promoter-driven GFP reporter transgene and provided by Dr. Shoen Kume) [17, 18] and ING112 (carrying a Ins1-promoter-driven GFP reporter transgene and provided by Dr. Shoen Kume) [19, 20].
15. Overconfluent culturing should be avoided because it leads to differentiation of the ESCs.
16. The coating of the synthetic fiber plate with PLL increase the amount of absorbed laminin to the plate [7].
17. The coating of the synthetic fiber plate with laminin facilitates the differentiation of mouse ESCs to pancreatic lineages [6, 7].
18. The total number of cells would be 5–10 $\times$ 10^6 from one 60 mm dish.
19. The cells should be resuspended in 1 mL of plating medium with a P1000 micropipette and then resuspended in a defined volume of the medium with a 10 mL pipette.
20. Multi-channel micropipette is convenient for plating the cells and changing the media.
21. The differentiation medium should be warmed at 37 °C in a water bath before use.

22. Add 10 μL of 20 μM recombinant proteins to 200 μL of cells to yield a final concentration of 1 μM.
23. The KRBH buffers should be warmed at 37 °C in a water bath before use.
24. This is a sample as low glucose. The buffer upon the cells should be carefully collected to avoid a contamination of cells.
25. This is a sample as high glucose. The buffer upon the cells should be carefully collected to avoid a contamination of cells.
26. This centrifugation step can remove the contaminated cells.
27. Alternative kit for mouse C-peptide is also available from ALPCO, etc.
28. BCA assay can be also used to detect the protein concentration.

References

1. Pagliuca FW, Millman JR, Gürtler M, Segel M, Van Dervort A, Ryu JH et al (2014) Generation of functional human pancreatic β cells in vitro. Cell 159:428–439
2. Rezania A, Bruin JE, Arora P, Rubin A, Batushansky I, Asadi A et al (2014) Reversal of diabetes with insulin-producing cells derived in vitro from human pluripotent stem cells. Nat Biotechnol 32:1121–1133
3. Nostro MC, Sarangi F, Yang C, Holland A, Elefanty AG, Stanley EG et al (2015) Efficient generation of NKX6-1+ pancreatic progenitors from multiple human pluripotent stem cell lines. Stem Cell Rep 4:591–604
4. Agulnick AD, Ambruzs DM, Moorman MA, Bhoumik A, Cesario RM, Payne JK et al (2015) Insulin-producing endocrine cells differentiated in vitro from human embryonic stem cells function in macroencapsulation devices in vivo. Stem Cells Transl Med 4:1214–1222
5. Sakano D, Shiraki N, Kikawa K, Yamazoe T, Kataoka M, Umeda K et al (2014) VMAT2 identified as a regulator of late-stage β-cell differentiation. Nat Chem Biol 10:141–148
6. Kaitsuka T, Noguchi H, Shiraki N, Kubo T, Wei FY, Hakim F et al (2014) Generation of functional insulin-producing cells from mouse embryonic stem cells through 804G cell-derived extracellular matrix and protein transduction of transcription factors. Stem Cells Transl Med 3:114–127
7. Kaitsuka T, Kojima R, Kawabe M, Noguchi H, Shiraki N, Kume S et al (2019) A culture substratum with net-like polyamide fibers promotes the differentiation of mouse and human pluripotent stem cells to insulin-producing cells. Biomed Mater. (in press) 14:045019
8. Joliot A, Prochiantz A (2004) Transduction peptides: from technology to physiology. Nat Cell Biol 6:189–196
9. Gump JM, Dowdy SF (2007) TAT transduction: the molecular mechanism and therapeutic prospects. Trends Mol Med 13:443–448
10. Vivès E, Schmidt J, Pèlegrin A (2008) Cell-penetrating and cell-targeting peptides in drug delivery. Biochim Biophys Acta 1786:126–138
11. Matsushita M, Tomizawa K, Moriwaki A, Li ST, Terada H, Matsui H (2001) A high-efficiency protein transduction system demonstrating the role of PKA in long-lasting long-term potentiation. J Neurosci 21:6000–6007
12. Matsui H, Tomizawa K, Lu YF, Matsushita M (2003) Protein therapy: in vivo protein transduction by polyarginine (11R) PTD and subcellular targeting delivery. Curr Protein Pept Sci 4:151–157
13. Noguchi H, Kaneto H, Weir GC, Bonner-Weir S (2003) PDX-1 protein containing its own antennapedia-like protein transduction domain can transduce pancreatic duct and islet cells. Diabetes 52:1732–1737
14. Noguchi H, Bonner-Weir S, Wei FY, Matsushita M, Matsumoto S (2005) BETA2/NeuroD protein can be transduced into cells due to an arginine- and lysine-rich sequence. Diabetes 54:2859–2866
15. Wilson ME, Scheel D, German MS (2003) Gene expression cascades in pancreatic development. Mech Dev 120:65–80

16. Hooper M, Hardy K, Handyside A, Hunter S, Monk M (1987) HPRT-deficient (Lesch-Nyhan) mouse embryos derived from germline colonization by cultured cells. Nature 326:292–295
17. Gu G, Wells JM, Dombkowski D, Preffer F, Aronow B, Melton DA (2004) Global expression analysis of gene regulatory pathways during endocrine pancreatic development. Development 131:165–179
18. Shiraki N, Yoshida T, Araki K, Umezawa A, Higuchi Y, Goto H et al (2008) Guided differentiation of embryonic stem cells into Pdx1-expressing regional-specific definitive endoderm. Stem Cells 26:874–885
19. Hara M, Wang X, Kawamura T, Bindokas VP, Dizon RF, Alcoser SY et al (2003) Transgenic mice with green fluorescent protein-labeled pancreatic beta -cells. Am J Physiol Endocrinol Metab 284:E177–E183
20. Higuchi Y, Shiraki N, Yamane K, Qin Z, Mochitate K, Araki K et al (2010) Synthesized basement membranes direct the differentiation of mouse embryonic stem cells into pancreatic lineages. J Cell Sci 123:2733–2742

Chapter 8

Cardiac Targeting Peptide: From Identification to Validation to Mechanism of Transduction

Kyle S. Feldman, Maria P. Pavlou, and Maliha Zahid

Abstract

Cell-penetrating peptides (CPPs), also known as protein transduction domains, were first identified 25 years ago. They are small, ~6–30 amino acid long, synthetic, or naturally occurring peptides, able to carry a variety of cargoes across the cellular membranes in an intact, functional form. These cargoes can range from other small peptides, full-length proteins, nucleic acids including RNA and DNA, nanoparticles, and viral particles as well as radioisotopes and other fluorescent probes for imaging purposes. However, this ability to enter all cell types indiscriminately, and even cross the blood–brain barrier, hinders their development into viable vectors. Hence, researchers have adopted various strategies ranging from pH activatable cargoes to using phage display to identify tissue-specific CPPs. Use of this phage display strategy has led to an ever-expanding number of tissue-specific CPPs. Using phage display, we identified a 12-amino acid, non-naturally occurring peptide that targets the heart with peak uptake at 15 min after a peripheral intravenous injection, that we termed Cardiac Targeting Peptide (CTP). In this chapter, we use CTP as an example to describe techniques for validation of cell-specific transduction as well as provide details on a technology to identify binding partner(s) for these ever-increasing plethora of tissue-specific peptides. Given the myriad cargoes CTP can deliver, as well as rapid uptake after an intravenous injection, it can be applied to deliver radioisotopes, miRNA, siRNA, peptides, and proteins of therapeutic potential for acute cardiac conditions like myocardial infarction, where the window of opportunity for salvaging at-risk myocardium is limited to 6 hrs.

Key words Cardiac targeting peptide, Protein transduction domains, Cell-penetrating peptides, Phage display, TriCEPS

1 Introduction

The cell plasma membrane is a semi-permeable barrier that is essential for cell integrity and survival but at the same time presents a barrier to delivery of cargoes. Hence, the ability of trans-activator of transcription (Tat) protein of the human immunodeficiency virus to enter cultured cells and promote viral gene expression was met with great enthusiasm [1, 2]. Shortly thereafter, Antennapedia homeodomain, a homeobox transcription factor of *Drosophila melanogaster*, was shown to enter nerve cells and regulate neural

Kumaran Narayanan (ed.), *Bio-Carrier Vectors: Methods and Protocols*, Methods in Molecular Biology, vol. 2211,
https://doi.org/10.1007/978-1-0716-0943-9_8,

morphogenesis [3]. Mapping of the domains responsible for this cell-penetrating ability led to the identification of two protein transduction domains, (also known as cell-penetrating peptides or CPPs); Tat corresponding to the 11 amino acid basic domain of HIV-1 Tat protein, and Penetratin corresponding to the 16 amino acid third helix of the Antennapedia domain. Subsequently, it was demonstrated that Tat fused to β-galactosidase and injected intraperitoneally in mice was internalized into multiple cell types including liver, heart, lung, kidney, and brain, delivering β-galactosidase in a functional form, highlighting the potential of Tat as a vector [4]. Currently, multiple cargoes in the form of peptides, proteins, nucleic acid, nanoparticles, and radioisotopes have been delivered using various CPPs [5, 6].

The ability of cationic or hydrophobic CPPs to transduce a wide variety of tissue types in vivo limits their utility because of lack of cell specificity. Phage display using libraries of various lengths and different bacteriophage strains have been utilized successfully to identify tissue-specific CPPs [7]. In our prior work, we identified a mildly basic, non-naturally occurring peptide (NH_2-A PWHLSSQYSRT-COOH) capable of specifically targeting normal cardiomyocytes (CMCs) in vivo in mice, which we hence termed Cardiac Targeting Peptide or CTP [7, 8]. Our detailed bio-distribution studies show that peak uptake occurs at 15 min with complete disappearance of fluorescently labeled CTP by 6 hrs [9].

Cellular responses to ligands such as peptides, proteins, pharmaceutical drugs, or entire pathogens are generally mediated through interactions with specific proteins expressed on the cell surface. The ligand-based receptor capture (LRC)-TriCEPS methodology has been designed to directly identify such ligand–receptor interactions under near-physiological conditions on living cells. The key component of the LRC methodology is a trifunctional cross-linker that combines the following moieties; a N-hydroxysuccinimide (NHS), a hydrazone, and a biotin or an azide [10, 11].

In a typical LRC-TriCEPS experiment, at least two treatment arms are performed in parallel: one with the ligand of interest and a second with a control ligand with a known target. The first step of the LRC-TriCEPS experiment includes the conjugation of the ligand to TriCEPS using the NHS-ester. The TriCEPS-ligand conjugates are then incubated with previously oxidized cells. During this phase, the hydrazone captures covalently the surface glycoproteins via the aldehydes that are introduced at the carbohydrates by mild oxidation (receptor capture). After the receptor-capture reaction, the cells are lysed and the azide or the biotin groups are used to purify the cross-linked proteins for downstream mass spectrometry (MS)-based analysis (cell lysis, protein enrichment, and digestion). Upon identification, the cell surface proteins in the ligand samples are compared to those in the control sample using label-free quantification (Fig. 1). Randomly identified cell surface proteins are expected to have equal abundance in both samples, whereas the corresponding targets are found enriched in the ligand sample [10, 11].

The LRC-TriCEPS methodology offers several unique advantages. It can be applied in a multitude of ligands ranging from small molecules to peptides, proteins, antibodies, and even whole viruses. Moreover, it does not require any genetic manipulation and therefore can be applied to a multitude of cell lines, including primary cells. Furthermore, targets are located within the context of the natural cell-specific surface microenvironment, so they are fully functional and exhibit their characteristic binding properties. Finally, it is hypothesis-free, meaning that no previous knowledge about the target is required.

A drawback of the methodology can be missing an interaction in case the target receptor is not expressed in the selected model system or TriCEPS coupling leads to hindering of the ligand–receptor interaction, which would interfere with ligand identification. In these scenarios, TriCEPS would not be useful. Additionally, the LRC-TriCEPS methodology enables the identification of only glycosylated targets; therefore, a small percentage of binding partners, less than 10%, cannot be identified.

In this chapter, we present details on the various methodologies that can be used to confirm tissue-specific internalization of a cell-specific CPP using CTP targeting the heart as an example. We also present details on LRC methodology.

2 Materials

2.1 Validation of Tissue-Specific CPPs

This section provides details on how to validate a candidate CPP for cell-specific transduction and contains protocols that can be modified for use in a variety of cell lines and organs, using fluorescently labeled peptides. We also provide a protocol for fluorophore conjugation of biotinylated CPPs. The methods detailed below use CTP in H9C2 cells, a rat cardiomyoblast cell line and wild-type mouse animal models as an example.

2.1.1 Transduction Assay Utilizing Fluorescence-Activated Cell Sorting or Confocal Imaging

1. Dulbecco's Modified Eagle's Medium—high glucose (DMEM).
2. Dulbecco's Phosphate Buffered Saline (PBS).
3. Heat-inactivated fetal bovine serum (FBS).
4. Antibiotic-Antimycotic.
5. Rat Cardiomyoblast Cell Line (H9C2; ATCC).
6. Rat Cardiomyoblast Cell Line (H9C2 Cells) Media: Add 50 mL of heat-inactivated FBS and 5 mL of antibiotic-antimycotic to 500 ml of DMEM. Filter the mixture and store at 4 °C.

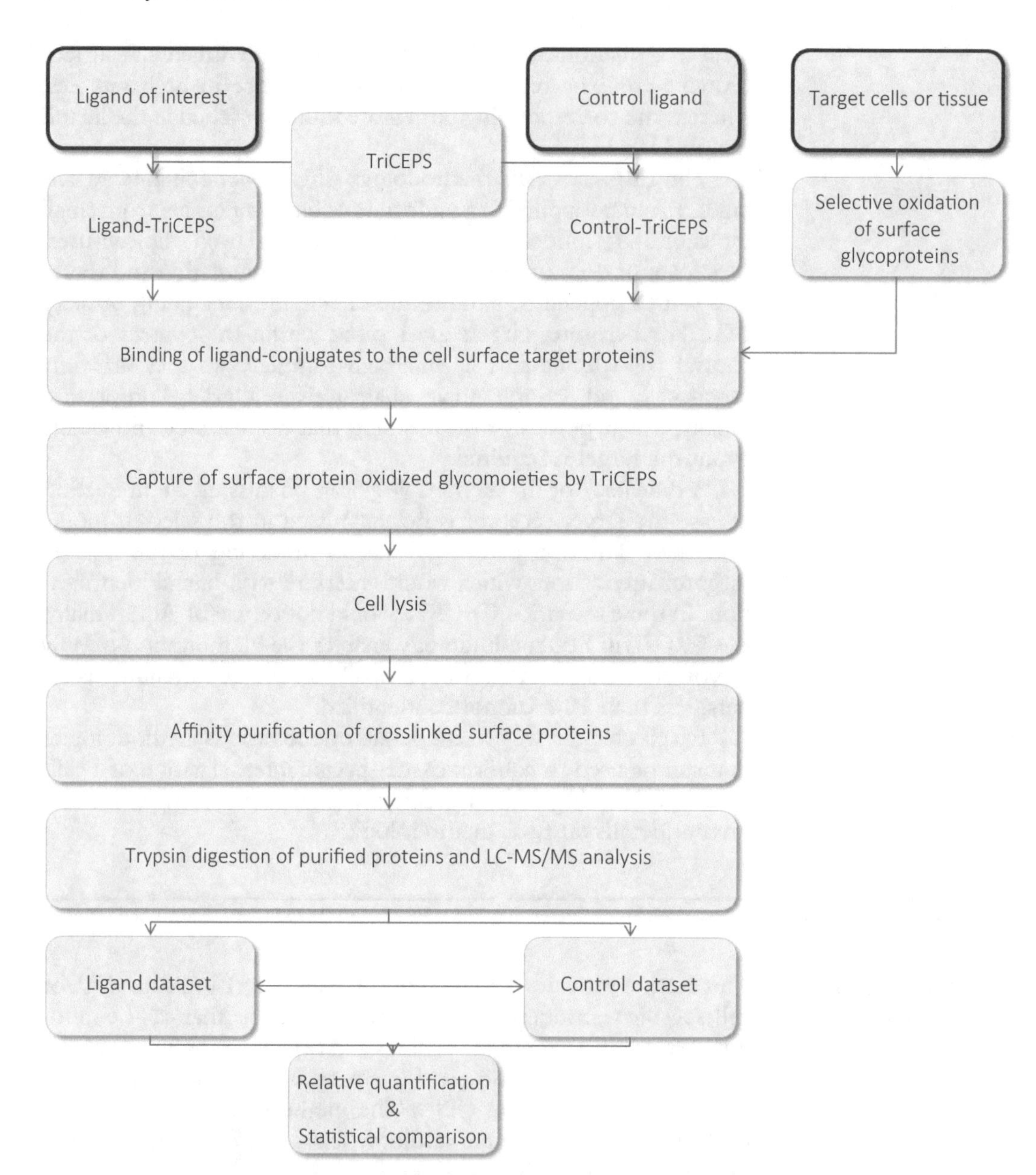

Fig. 1 Schematic of the workflow using LRC-TriCEPS technology. In a typical LRC-TriCEPS experiment, at least two treatment arms are performed in parallel: one with the ligand of interest and a second with a control ligand (i.e., a ligand with a known target). The first step of the LRC-TriCEPS experiment includes the conjugation of the two ligands to TriCEPS (ligand coupling). The TriCEPS-ligand conjugates are then incubated with previously oxidized cells under near-physiological conditions. During this phase, transient and stable ligand–receptor interactions will result in covalent capture events between TRICEPS and nearby carbohydrates (receptor capture). After the receptor-capture reaction, the cells are lysed, and cross-linked proteins are isolated and processed for mass spectrometry-based analysis (cell lysis, protein enrichment and digestion). Upon identification, the relative abundance of cell surface proteins in the ligand samples are compared to those in the control sample using label-free quantification. Randomly identified cell surface proteins are expected to have equal abundance in both samples, whereas the corresponding receptors are found enriched in the ligand sample

7. Solution of trypsin (0.25%) and ethylenediamine tetraacetic acid (EDTA).
8. Immortal Mouse Fibroblast Cell Line (3T3; ATCC).
9. 16% Paraformaldehyde.
10. FACS Fixation Buffer: 10 mL of 16% PFA diluted with 30 mL of PBS. Solution should be stored with light protection at 4 °C.
11. Live/Dead Fixable Aqua Dead Cell Stain, 405 nm excitation. This dye is reconstituted in DMSO and is stored at −20 °C, light protected, when not in use.
12. Dimethyl Sulfoxide (DMSO).
13. Cy5.5-CTP-amide: Prepared by peptide synthesis, conjugated with Cy5.5 fluorophore, and purified using HPLC. Lyophilized powder is reconstituted in DMSO at 10 mM concentration. After reconstituting, store at -80 °C.
14. Glass-bottomed dishes (MatTek).

2.1.2 Bio-distribution In Vivo

1. 6- to 12-week-old albino, female mice (CD1, Charles River).
2. 28G 0.5 mL Insulin syringes.
3. 100 mg/mL Ketamine HCl (KetaVed).
4. 20 mg/mL Xylazine (AnaSed).
5. Ketamine/Xylazine Solution: Ketamine HCl is mixed with Xylazine 1:1 to produce a stock solution containing 50 mg/mL ketamine and 10 mg/mL Xylazine. This solution is made fresh before each use.
6. 26G Needles.
7. 3 mL Syringe.
8. 10% Buffered Formalin Phosphate.
9. Paraplast X-TRA (SIGMA).
10. Ethanol.
11. Xylene.
12. 10× Tris Buffered Saline (TBS).
13. 1× TBS Solution: 10× TBS is diluted 1:10 with deionized water. This solution can be stored at room temperature.
14. Microscope Slides.
15. Cover Glass.
16. Dapi Fluoromount G (SouthernBiotech).
17. Sparkle Optical Lens Cleaner.
18. Tissue-Tek Processing/Embedding Cassettes (ThermoFisher).

2.2 LRC-TriCEPS Methodology for Identifying Binding Partners

This section provides details on the materials needed to perform an LRC-TriCEPS experiment.

1. 5-methoxyanthranilic acid (Sigma-Aldrich).
2. Acetonitrile LC-MS grade (Fisher Scientific).
3. Alkyne agarose.
4. Ammonium bicarbonate.
5. Calcium chloride ($CaCl_2$).
6. Copper(II)sulfate ($CuSO_4$.
7. Ethylenediaminetetraacetic acid (EDTA).
8. Formic acid, 98–100% (vol/vol).
9. Glycine.
10. HEPES pH 8.2, 1 M buffer solution.
11. Iodoacetamide (Sigma-Aldrich).
12. Isopropanol for molecular biology, $\geq$99.5%.
13. Mobicol spin columns (Life Systems Design).
14. Mobicol filter 35 μm pore size (Life Systems Design).
15. PBS, pH 7.4.
16. Phosphoric acid 85 wt.% (H_3PO_4).
17. Protease inhibitors cOmplete (Sigma-Aldrich).
18. Rapigest SF (waters).
19. Sequencing grade modified trypsin (Promega).
20. Sodium ascorbate.
21. Sodium bicarbonate.
22. Sodium chloride.
23. Sodium dodecyl sulfate (SDS, Sigma-Aldrich).
24. Sodium hydroxide solution.
25. Sodium-metaperiodate ($NaIO_4$, Sigma-Aldrich).
26. TriCEPS v.3.0 (Dualsystems Biotech).
27. Tris (2-carboxyethyl)phosphine hydrochloride (TCEP, Sigma-Aldrich).
28. Tris (3-hydroxypropyltriazolylmethyl)amine (THPTA, Sigma-Aldrich).
29. Tris (hydroxymethyl)aminomethane base.
30. Urea.
31. Water HPLC gradient grade.

2.2.1 Buffer Preparation

1. Ligand coupling: HEPES 25 mM pH 8.2: Dilute 1 M HEPES with HPLC water and adjust the pH to 8.2 with NaOH.

2. Receptor capture: PBS pH 6.5: Acidify 500 mL of PBS (pH 7.4) by adding 105 μL of 85% (vol/vol) phosphoric acid.
3. Sodium (meta) periodate 1.5 mM: Dissolve 16 mg of sodium (meta) periodate in 50 mL of PBS pH 6.5 and vortex the solution until it is fully dissolved (always prepare fresh, protect from light).
4. 5-Methoxyanthranilic acid 5 mM: Dissolve 84 mg of 5-methoxyanthranilic acid in 100 mL PBS pH 7.4; stir with heating at 40 °C until it is fully dissolved and adjust pH to 7.4 with NaOH (protect from light).
5. Lysis buffer: Dissolve 4.8 g urea to 10 mL deionized water, stir until it is fully dissolved. Add 0.1% (vol/vol) Rapigest SF and one Protease Inhibitors cOmplete tablet, stir until it is fully dissolved (protect from light).
6. Protein enrichment: $CuSO_4$ 100 mM: Dissolve 25 mg of CuSO4 to 1 mL deionized water.
7. THPTA 100 mM: Dissolve 44 mg of THTPA to 1 mL deionized water (store aliquots in −20 °C).
8. Sodium ascorbate 1 M: Dissolve 200 mg of sodium ascorbate in 1 mL of deionized water (prepare fresh).
9. Click Chemistry Master Mix (2×): In 835 μL deionized water, add 125 μL THPTA, 20 μL CuSO4, and vortex; then add 20 μL sodium ascorbate and vortex again.
10. Tris-HCl 1 M: Dissolve 121.1 g of Tris base in 800 mL of deionized water; adjust the pH to 8.0 with HCl.
11. NaCl 5 M: Dissolve 292gm of NaCl in 1 L of deionized water.
12. EDTA 100 mM: Dissolve 1.86 g of EDTA to 50 mL of deionized water and adjust pH to 7.4 with NaOH.
13. SDS wash buffer (1% SDS, 100 mM Tris, 250 mM NaCl, 5 mM EDTA, pH 8): Mix 10 mL 1 M Tris–HCL, 5 mL NaCl, 5 mL 100 mM EDTA, and 1gm of SDS in a final volume of 100 mL.
14. Urea wash buffer (8 M urea in 100 mM Tris pH 8): Mix 48 g urea and 10 mL 1 M Tris to a final volume of 100 mL.
15. Isopropanol 80%: Mix 80 mL of isopropanol and 20 mL deionized water.
16. $NaHCO_3$ 100 mM: Dissolve 84 mg sodium bicarbonate in 100 mL deionized water, and adjust the pH to 11.0 with NaOH.
17. Acetonitrile 20%: Mix 20 mL of acetonitrile and 80 mL of deionized water.
18. Ammonium bicarbonate 50 mM: Dissolve 39 mg of ammonium bicarbonate in 10 mL of deionized water.

19. $CaCl_2$ 50 mM: Dissolve 55.5 mg in 10 mL deionized water.
20. Digestion buffer: Mix 10 mL 1 M Tris, 4 mL 50 mM $CaCl_2$, 10 mL acetonitrile to a total volume of 100 mL.
21. Formic acid 10%: Mix 10 mL of formic acid and 90 mL HPLC water.
22. C18 wash buffer: Mix 80 mL acetonitrile, 1 mL 10% Formic acid and 19 mL HPLC water.
23. C18 loading buffer: Mix 2 mL acetonitrile, 1 mL 10% formic acid, and 19 mL HPLC water.
24. C18 elution buffer: Combine 50 mL acetonitrile, 1 mL 10% formic acid, and 49 mL HPLC water.

3 Methods

3.1 Transduction Assay Using Fluorescence-Activated Cell Sorting

This protocol details how to perform an in vitro transduction assay of a candidate CPP for the purposes of validation. Here we provide a protocol to validate the specific transduction of Cy5.5-CTP into H9C2 cells with 3T3 mouse fibroblasts as a negative control.

1. Thaw cells from −180 °C liquid N_2 storage and plate onto a T-75 tissue culture treated flask with vented caps. Passage H9C2 cells for a minimum of three passages, trypsinizing at 60–70% confluence and plating at a 1:3 to 1:5 dilution (one 70% confluent T-75 flask yields ~one million cells) (*see* **Note 1**).
2. Passage the 3T3 cells for a minimum of 3 passages (similar to details given above for H9C2 cells), after thawing from −180 °C (*see* **Note 1**).
3. Once cells are 70% confluent, aspirate media, wash once with pre-warmed PBS, aspirate PBS and add 2 mL of Trypsin/EDTA, incubate for 5–7 min until cells have rounded up and are coming off in sheets.
4. Neutralize with 10–12 mL of pre-warmed media, collect cells, spin at 300 × *g* for 6 min to pellet the cells.
5. Aspirate supernatant and wash cell pellet twice with 10–12 mL of pre-warmed media.
6. Count cells using a hematocytometer and plate at a density of 1×10^5 cells/well of a 12-well plate.
7. Twenty-four hours post-plating, aspirate and replace with pre-warmed 1 mL of media to which 10 μM of Cy5.5-CTP final concentration and 1 μL/mL of Live/Dead Fixable Aqua Dead Cell Stain has been added (*see* **Notes 2** and **3**). Return to incubator for 30 min. All cell works/groups/treatments are done in triplicate.

8. Suggested groups are H9C2/3T3 cells with no treatment or random peptide or CTP with Live/Dead stain added to all groups.
9. After the incubation period, wash cells extensively (at least 3×) with pre-warmed PBS, trypsinize, and centrifuge to collect cell pellet.
10. Wash the cell pellet with media once.
11. Aspirate the supernatant media above the cell pellet and resuspend in 1 mL of FACS Fixation Buffer.
12. Incubate the cells for 10 min at room temperature.
13. Add up to 5 mL of PBS and centrifuge to collect cell pellet.
14. Aspirate the supernatant and resuspend the cells in 200 μL of PBS.
15. Store the resuspended cells on ice, light protected.
16. In the flow cytometer run bleach and distilled water for 1 min each to clear the lines. Set to standby with a tube of distilled water.
17. Set up a protocol, specifying number of events to log and gates. Gates to set up would be to select for cells using FSC-A vs. SSC-A graph, a gate to select for live cells excluding the Live/Dead Fixable Aqua Dead Cell Stain using an SSC-A vs. 405 nm violet laser 535/50 filter, and graph to display SSC-A vs. 640 nm red laser 660/20 filter.
18. Vortex the cell sample, insert into the cytometer and adjust voltages so that the 660/20 fluorescence peaks in the middle of the graph. Apply these settings to the rest of the samples.

3.2 Transduction Assay Using Confocal Microscopy

This protocol is to acquire qualitative data about the transduction of a CPP using confocal microscopy. Here we utilize H9C2 cells incubated with Cy5.5 labeled CTP with 3T3 cells as a negative control. This protocol can be modified for use in ther cell lines, peptides, and fluorophores and can be modified for simultaneous fluorescent labeling of different cellular organelles for co-localization of CPP to particular cellular compartments.

1. Passage H9C2 cells and 3T3 cells as detailed in Subheading 3.1 above and plate in an optical glass-bottomed dish at a cell density of 5×10^4 cells/well.
2. Twenty-four hours post-plating, aspirate, and replace with pre-warmed media to which Cy5.5-CTP has been added to give a final concentration of 10 μM.
3. Separate group of cells should be treated with vehicle (PBS) only and a random peptide similarly labeled with Cy5.5.

4. Return cells to incubator for 30 min. After the incubation period, wash cell extensively (at least 3×) with pre-warmed media. The cells are now ready to be imaged using confocal microscopy.

3.3 Bio-distribution In Vivo

This protocol details how to perform bio-distribution studies for a candidate CPP. Here we provide an example of bio-distribution studies using Cy5.5-CTP in a mouse model. A number of organs should be harvested, such as the heart, lung, liver, kidney, brain, spleen, stomach, large intestine, small intestine, skeletal muscle, bone, and testes/ovaries, with each time point performed in triplicate.

1. Weigh mice and anesthetize with Ketamine/Xylazine (2 μL/g of tissue weight) administered intramuscularly or intraperitoneally. Adequate level of anesthesia will be achieved in 5–7 min, as assessed by lack of response to toe pinch.
2. Calculate a 10 mg/kg dose of Cy5.5-CTP, dilute to no more than 200 μL, and inject either retro-orbitally or through tail vein injection using an insulin syringe (*see* **Note 2**).
3. Allow peptide to circulate for the pre-specified time.
4. Euthanize mouse using Institutional Animal Care and Use Committee's specified method and open the chest cavity.
5. Place a nick in the right atrium and, using a 26G needle, inject 3 mL of 10% Buffered Formalin Phosphate for the dual purpose of perfusion fixing the organs of the mouse and flushing out red blood cells.
6. Dissect out the organs of interest and store each organ individually in 10% Buffered Formalin Phosphate in a volume at least 20 times the volume of the tissue, weight per volume, for a minimum of 48 hrs, light protected, at room temperature (*see* **Note 4**).
7. Transfer organs into Tissue-Tek Processing/Embedding Cassettes and process the organs using a Tissue-Tek VIP processing machine.
8. Dehydrate the tissue in 70% EtOH for 30 min, followed by 80% EtOH for 30 min, 95% EtOH for 30 min, 95% EtOH for 30 min, 100% EtOH for 15 min, 100% EtOH for 20 min, and finally 100% EtOH for 20 min.
9. Clear the tissue in xylene twice, 30 min for each xylene treatment.
10. Infiltrate the cleared tissue with paraffin wax four times at 60 °C for 30 min with each treatment.
11. Embed in paraffin using metal molds. Fill the mold with molten paraffin kept at 65 °C and transfer to a cold plate. As the paraffin at the bottom of the mold begins to solidify, place the

organ in the desired orientation. Place a labeled cassette on top of the mold as a backing and overfill with molten paraffin. Allow to cool until completely solid.

12. The block can then be stored in a -20 °C freezer over night to completely solidify.
13. Prepare a 38 °C water bath with distilled water.
14. Set up the microtome with a blade angle of 6° and a section thickness of 15 μm.
15. Section the tissue in the microtome by cutting the desired plane and then placing the blocks face down in the water back for 5 min or until the tissue has absorbed some moisture.
16. Place the tissue onto a flat ice block for 10 min.
17. Place a fresh blade onto the microtome and cut sections with a thickness of 8 μm. Discard bad paraffin ribbons until a ribbon of sufficient length and quality is produced.
18. Quality ribbons are then picked up with forceps and floated on the surface of the 38 °C water bath. Let the sections sit on the surface until they smooth out, taking care to not leave them too long to prevent the paraffin from disintegrating and tearing apart the section.
19. Float the flattened sections onto the surface of clean glass slides.
20. Place the slides into a 65 °C oven for 30 min to melt the wax. These slides can be stored at room temperature with light protection.
21. Deparaffinize the slides in Xylene three times, 10 min for each treatment.
22. Rehydrate the tissue in 100% EtOH for 5 min, followed by 95% EtOH for 5 min, 70% EtOH for 5 min, 50% EtOH for 5 min, and finally 1× TBS for 5 min.
23. Mount the slides with coverslips using 125 μL of Dapi Fluoromount G.
24. Dry slides overnight at room temperature, light protected.
25. Image slides using confocal microscopy.

3.4 Ligand-Receptor Coupling Experimental Protocol for TriCEPS

This section provides details on how to perform an LRC-TriCEPS experiment and contains protocols that can be modified for use in a variety of cell lines and ligands of interest.

3.4.1 Ligand Coupling

The following protocol (Fig. 1) describes step-be-step an LRC experiment using TriCEPS v.3.0 (azide-containing) for generating one ligand and one control sample; therefore, the quantities need to be multiplied by the number of desired biological replicates (minimum three recommended) (*see* **Notes 14–18**).

1. Dissolve 100 μg of the ligand of interest in 50 μL of 25 mM HEPES (pH 8.2) in an Eppendorf tube (ligand sample).
2. Dissolve 100 μg of the control ligand in 50 μL of 25 mM HEPES (pH 8.2) in an Eppendorf tube (control sample).
3. If no control ligand is available, quench TRICEPS with 100 μg of glycine in 50 μL of 25 mM HEPES (pH 8.2).
4. Add 0.5 μL of the TRICEPS solution to each of the ligand and control sample and mix them thoroughly by pipetting.
5. Incubate the mixture at room temperature under constant gentle agitation for 90 min.

3.4.2 Receptor Capture (All Steps Performed at 4 °C)

1. Collect 2×10^7 cells with gentle scraping (for adherent cells) or centrifugation (for suspension cells) (*see* **Notes 19** and **20**).
2. Wash cells once with 50 mL PBS pH 6.5 in a 50 mL centrifuge tube.
3. Centrifuge the cells at 300 × *g* for 5 min at 4 °C, discard the supernatant, and resuspend the cells in the sodium (meta) periodate buffer.
4. Incubate cells at 4 °C in the dark for 15 min, under constant gentle agitation.
5. Centrifuge the cells at 300 × *g* for 5 min at 4 °C, discard the supernatant, and wash cells with 50 mL of PBS pH 7.4.
6. Centrifuge cells at 300 × *g* for 5 min at 4 °C and resuspend cells in 20 mL 5-methoxyanthranilic acid buffer.
7. Split the cell suspension into two 15 mL centrifuge tubes.
8. Add the TRICEPS-coupled ligand of interest to one of the tubes and the coupled control ligand to the other.
9. Incubate ligands with cells for 90 min at 4 °C, under constant gentle agitation.
10. Centrifuge the cells at 300 × *g* for 5 min at 4 °C and wash cells with 10 mL PBS pH 7.4.
11. Centrifuge the cells at 300 × *g* for 5 min at 4 °C, remove the supernatant, and freeze the cell pellets.

3.4.3 Cell Lysis and Protein Enrichment

1. Resuspend the cell pellets in 800 μL of lysis buffer and transfer the lysates to two Eppendorf tubes.
2. Sonicate the lysates using three 30 s sonication pulses and remove debris by centrifugation at 16,000 × *g* for 10 min.
3. Wash 200 μL alkyne agarose (per replicate) with 1.8 mL deionized water and add to the lysates.
4. Add 1 mL of 2× click chemistry mastermix and incubate the samples for 18 h (±2 h) at room temperature under gentle agitation.

5. Pellet agarose beads by centrifugation for 4 min at 300 × *g*, remove supernatant, and wash beads with 1.8 mL deionized water.
6. Resuspend beads in 1 mL SDS wash buffer and reduce bead-bound proteins with 5 mM TCEP for 15 min at 55 °C and 15 min at room temperature.
7. Pellet beads at 300 × *g* for 4 min to remove supernatant.
8. Alkylate the proteins with 40 mM iodoacetamide for 30 min, at room temperature, in the dark.
9. Transfer beads to Mobicol classic 35-μm filters and wash beads with 10 mL of the following buffers: SDS wash buffer, urea wash buffer, 5 M NaCl, 80% isopropanol, 100 mM $NaHCO_3$, 50 mM ammonium bicarbonate (60 °C), and 20% acetonitrile.
10. Transfer the beads in fresh Mobicol tube.

3.4.4 On-Bead Trypsin Digestion

1. Resuspend the beads in 400 μL digestion buffer.
2. Add 1 μg sequencing grade modified trypsin and incubate for 16 h at 37 °C.
3. Collect the peptides, wash beads twice with 50 mM ammonium bicarbonate.
4. Acidify samples with 10% formic acid to pH 3.

3.4.5 Peptide Purification

1. Desalt the peptides using UltraMicroSpin C18 Columns with 5–60 μg capacity for tryptic peptide fraction according to manufacturer's instructions.
2. Dry the eluted peptides in a SpeedVac and store them at −80 °C until further analysis.

3.4.6 LC-MS/MS and Data Analysis

LRC samples are of medium-to-high complexity and need to be analyzed with a highly sensitive, high mass accuracy mass spectrometer. Analyze peptides with a standard shotgun mass spectrometry-based workflow and perform label-free quantification to extract relative protein abundance.

3.4.7 Statistical Analysis

Perform statistical analysis to calculate protein fold changes and their statistical significance between paired conditions; numerous free software packages exist for statistical analysis such as MSstats [12] and SafeQuant [13].

3.4.8 Data Visualization and Interpretation

A volcano plot combines a measure of statistical significance from a statistical test with the magnitude of the change, enabling quick visual identification of proteins that were significantly enriched in the ligand of interest samples. The *x*-axis represents the mean ratio fold change (on a log2 scale). The *y*-axis represents the statistical significance *p*-value of the ratio fold change for each protein (on a

−log10 scale). Proteins that are enriched in one of the samples will plot either left or right of the *x*-axis origin, indicating in which sample the target protein is enriched.

4 Notes

1. H9C2 and 3T3 cells should be passaged once they are about 70% confluent. Do not allow to grow to complete confluency as they will begin to differentiate.
2. Avoid freeze-thaw cycles of peptides once the lyophilized powder is in solution in DMSO. Aliquot into amber/black Eppendorf tubes and store long term (6–24 months) at −80 °C. Lyophilized powder can be stored longer term in −20 °C, as long as it is light protected, for fluorescently labeled CPPs.
3. Similarly, avoid freeze-thaw cycles of the Live/Dead stain once reconstituted into DMSO.
4. Organs can be stored long term in formalin, or after being in formalin for 48 h, can be transferred into 70% EtOH for long-term storage, with light protection for both.
5. Tissue processing machines can be programmed to perform the series of solution exchanges needed to process tissue automatically though hand processing is a perfectly acceptable alternative provided the samples are protected from light.
6. Tissue blocks can be popped out of the molds and stored at room temperature for years with light protection.
7. Do not oversoak tissue until it is swollen. Soaking until the edges of the tissue lighten is sufficient.
8. When selecting fluorophores for tracking a CPP, care should be taken to minimize spectra overlap, which can be an issue where multiple fluorophores are needed. Another factor in selecting fluorophores is the system being used in the experiment.
9. Minimize tissue autofluorescence and possible false-positive results by choosing fluorophores in the red or far-red spectra.
10. Avoid the dye TAMRA, which can produce very high background due to membrane association.
11. When using multiple cell types in flow cytometry; differences in cell size may need to be corrected for to ensure the results of the experiment are valid. This can be addressed by dividing the measured fluorescence by the measured forward scatter, a measure of cell size. Doing this would produce a measure of the density of fluorescence per cell. This method could be used to compare the results from multiple cell types.

12. When using confocal microscopy such as confocal, avoid saturated images, which are not quantitatively useful, by adjusting the saturation and gain settings.
13. Bleaching of a sample should be avoided.
14. TRICEPS coupling is favored under alkaline conditions and the coupling buffer must not contain primary amines (e.g., Tris). The recommended buffer is 25 mM HEPES pH 8.2.
15. It is also recommended to use 50 μg of TriCEPS with a high TriCEPS:ligand ratio to increase the number of carbohydrate structures that can be captured per ligand; coupling ratio of 50 μg of TRICEPS per 100 μg of ligand is recommended.
16. The coupling reaction is competed by the hydrolysis of NHS-ester and given that the hydrolysis occurs more readily in dilute protein solutions, a more concentrated solution is recommended. A coupling reaction with 100 μg of ligand per 50–100 μL of buffer is proposed.
17. The required ligand amount is roughly 100 μg per replicate and the ligand buffer should not contain primary amines; in case the ligand buffer is not compatible, the ligand should be dialyzed against 25 mM HEPES pH 8.2.
18. Given that coupling of the cross-linker may affect the bioactivity or binding properties of the ligand, binding and functional assays using the ligand-TriCEPS conjugates prior to the LRC experiments are recommended.
19. During collection of cells, no reagents containing proteases can be used as they will result in digestion of the extracellular proteins.
20. A total number of approximately 10^7 cells per sample are recommended; however, different cell types exhibit different cell size and surface. As a rule of thumb, the cell pellet of each sample at the end of the LRC should be between ca. 40 and 100 μL to render a sufficient protein concentration for further analysis.

Acknowledgments

M.Z. and K.S.F. are supported by American Heart Association Scientist Development Award 17SDG33411180, and by a grant awarded under the Pitt Innovation Challenge (PinCh) through the Clinical and Translational Science Institute of the University of Pittsburgh, through National Institutes of Health, UL1TR001857.

Disclosures: M.Z. along with Paul D. Robbins (Professor, University of Minnesota, Minnesota, MN, USA) hold a patent on the use of cardiac targeting peptide as a cardiac vector (Cardiac-specific protein targeting domain, U.S. Patent Serial No. 9,249,184). M.Z. also serves as Chief Scientific Officer and on the Board of Directors of the startup Vivasc Therapeutics Inc., and holds substantial equity in it. M.P. is the Chief Scientific Officer of Dualsystems Biotech AG that holds an exclusive license for the LRC technology as covered in patent application WO2012/104051.

References

1. Green M, Loewenstein PM (1988) Autonomous functional domains of chemically synthesized human immunodeficiency virus tat transactivator protein. Cell 55:1179–1188
2. Frankel AD, Pabo CO (1988) Cellular uptake of the tat protein from human immunodeficiency virus. Cell 55:1189–1193
3. Joliot A, Pernelle C, Deagostini-Bazin H, Prochiantz A (1991) Antennapedia homeobox peptide regulates neural morphogenesis. Proc Natl Acad Sci U S A 88:1864–1868
4. Schwarze SR, Ho A, Vocero-Akbani A, Dowdy SF (1999) In vivo protein transduction: delivery of a biologically active protein into the mouse. Science 285:1569–1572
5. Zahid M, Robbins PD (2015) Cell-type specific penetrating peptides: therapeutic promises and challenges. Molecules 20:13055–13070
6. Ramsey JD, Flynn NH (2015) Cell-penetrating peptides transport therapeutics into cells. Pharmacol Ther 154:78–86
7. Zahid M, Robbins PD (2011) Identification and characterization of tissue-specific protein transduction domains using peptide phage display. Methods Mol Biol 683:277–289
8. Zahid M, Phillips BE, Albers SM, Giannoukakis N, Watkins SC, Robbins PD (2010) Identification of a cardiac specific protein transduction domain by in vivo biopanning using a M13 phage peptide display library in mice. PLoS One 5:e12252
9. Zahid M, Feldman KS, Garcia-Borrero G et al (2018) Cardiac targeting peptide, a novel cardiac vector: studies in bio-distribution, imaging application, and mechanism of transduction. Biomol Ther 8(4):147
10. Frei AP, Moest H, Novy K, Wollscheid B (2013) Ligand-based receptor identification on living cells and tissues using TRICEPS. Nat Protoc 8:1321–1336
11. Sobotzki N, Schafroth MA, Rudnicka A et al (2018) HATRIC-based identification of receptors for orphan ligands. Nat Commun 9:1519
12. Choi M, Chang CY, Clough T et al (2014) MSstats: an R package for statistical analysis of quantitative mass spectrometry-based proteomic experiments. Bioinformatics 30:2524–2526
13. Timo Glatter T, Christina Ludwig C, Ahrné E et al (2012) Large-scale quantitative assessment of different in-solution protein digestion protocols reveals superior cleavage efficiency of tandem Lys-C/trypsin proteolysis over trypsin digestion. J Proteome Res 11:5145–5156

Chapter 9

Lactoferricin-Derived L5a Cell-Penetrating Peptide for Delivery of DNA into Cells

Natalie J. Holl, Moumita Dey, Yue-Wern Huang, Shiow-Her Chiou, and Han-Jung Lee

Abstract

Cell-penetrating peptides (CPPs) are small peptides which help intracellular delivery of functional macromolecules, including DNAs, RNAs, and proteins, across the cell membrane and into the cytosol, and even into the nucleus in some cases. Delivery of macromolecules can facilitate transfection, aid in gene therapy and transgenesis, and alter gene expression. L5a (RRWQW), originally derived from bovine lactoferricin, is one kind of CPPs which can promote cellular uptake of plasmid DNA and enters cells via direct membrane translocation. The peptide complexes noncovalently with DNA over a short incubation period. DNA plasmid and L5a complex stability is confirmed by a decrease in mobility in a gel retardation assay, and successful transfection is proven by the detection of a reporter gene in cells using fluorescent microscopy. Here, we describe methods to study noncovalent interactions between L5a and plasmid DNA, and the delivery of L5a/DNA complexes into cells. L5a is the one of the smallest CPPs discovered to date, providing a small delivery vehicle for macromolecules in mammalian cells. A small vehicle which can enter the nucleus is ideal for efficient gene uptake, transfer, and therapy. It is simple to complex with DNA plasmids, and its nature allows mammalian cells to be easily transfected.

Key words Cell-penetrating peptide, Electrophoretic mobility shift assay, Fluorescent microscopy, Gel retardation, Gene therapy, Intracellular delivery, N/P ratio, Protein transduction domain, Transgenesis

1 Introduction

The internalization of exogenous materials into cells through the plasma membrane is an important function of all eukaryotic cells. The use of cell-penetrating peptides (CPPs) has recently gained attention as a bright field to deliver therapeutic and diagnostic molecules into the cell in a nontoxic manner [1]. CPPs are peptides that are able to translocate across cellular membranes and to deliver diverse cargos into cells [2]. Most synthetic or natural CPPs are shorter than 30 amino acids [3, 4]. In 1988, the first CPP, trans-activator of transcription (TAT) of human immunodeficiency virus,

Kumaran Narayanan (ed.), *Bio-Carrier Vectors: Methods and Protocols*, Methods in Molecular Biology, vol. 2211, https://doi.org/10.1007/978-1-0716-0943-9_9, © Springer Science+Business Media, LLC, part of Springer Nature 2021

was introduced and commonly used [5, 6]. An increasing number of CPPs have been subsequently identified and extensively studied over the past three decades. Currently, about 1850 CPP sequences have been deposited in the CPPsite 2.0 database [7].

CPPs are capable of penetrating the cell membrane due to cationic, amphipathic, or hydrophobic domains [1, 8–11]. They are taken up in almost all cell types [1]. The CPP family hosts a wide array of variously structured proteins which differ in charge and polarity [3]. Although most are linear, some have α-helix and β-sheet secondary structure [3]. CPP uptake is often mediated by endocytosis [4], and it is common for CPPs to have stretches of polyarginine [3, 12]. In fact, a simple polyarginine sequence (such as SR9, RRRRRRRRR) can act as a CPP [11].

Transfection and gene expression can be controlled by the delivery of macromolecules for gene therapy. CPP carriers can bring transcription factors [13], RNA interference (RNAi) [1, 11], plasmid DNA [1, 11], functional proteins [1, 11], and other nanomaterials [11] into cells. Some CPPs are specifically designed to naturally enter the nucleus to deliver cargo with nuclear localization sequences and other motifs [4]. Delivery of these compounds can directly alter or influence the transcription of genes, or add new genes into cells.

Recently, several CPPs have been identified from bovine lactoferricin [8]. Among them, the novel pentapeptide L5a (RRWQW) has demonstrated the ability to form stable complexes with plasmid DNA and to transfer plasmid DNA into cells with no cytotoxicity [8]. L5a represents one of the shortest CPPs characterized to date and has helical structure [9]. Mechanistic studies have revealed that L5a appears to enter A549 cells by direct membrane translocation [9], while L6 (RRWQWR) and HL6 (CHHHHHRRWQWRHH HHHC), also derived from bovine lactoferricin, are internalized by endocytosis [10] and direct membrane translocation [14], respectively. Lactoferricin-derived CPPs should not be preferential to cells and will enter most cells effectively given their ability to pass through cell membranes. All three lactoferricin-derived CPPs (L5a, L6, and HL6) are positively charged, allowing electrostatic complexing with negatively charged macromolecules, like DNA and carboxyl-functionalized nanoparticles [8, 14].

The use of a small peptide derived from lactoferricin has several advantages. Derivation from mammalian proteins decreases the chance for immunogenicity compared to peptides from the transduction domain of viral proteins as our bodies have defenses in place for disease-causing proteins [8]. Although the size of CPPs does not affect transduction efficiency when complexed with DNA [9], a smaller delivery vehicle is advantageous. Small peptides have uniform and rapid biodistribution which reaches deep into tissues [15]. The short length of L5a also makes it easy to synthesize artificially [9, 15].

In this chapter, we describe protocols to (1) assess noncovalent interactions between L5a and plasmid DNA and (2) study the cellular internalization of L5a/DNA complexes as a safe tool for cellular transgenesis. This system is an effective method for DNA delivery as it is nontoxic and simple to complex with DNA. The potential applications may include, but not limited to, gene therapy, transgenesis, molecular interactions, and identification of signaling pathway and molecular mechanisms.

2 Materials

2.1 General Molecular and Cell Biology Reagents

1. Synthetic L5a peptide: Synthesize L5a (RRWQW) from a custom *peptide synthesis service provider*. Store at −20 °C.
2. Plasmid DNA: The pEGFP-N1 plasmid consists of the enhanced green fluorescent protein (*EGFP*) coding sequence under the control of the immediate early *cytomegalovirus* (*CMV*) promoter. Store at −20 °C. Amplify the plasmid using a bacterial strain of choice (like DH5α) and obtain 800 μg of purified plasmid DNA using a Plasmid DNA Maxiprep Purification kit.
3. Cell culture medium: Roswell Park Memorial Institute (RPMI) 1640 medium supplemented with 10% fetal bovine serum (FBS) and 1× antibiotic-antimycotic solution, such as penicillin, streptomycin, and amphotericin B. Add 50 mL FBS and 5 mL 100× antibiotic-antimycotic solution into 445 mL RPMI 1640 medium. Store at 4 °C.
4. Phosphate buffered saline (PBS): 137 mM NaCl, 2.7 mM KCl, 10 mM Na_2HPO_4, and 1.8 mM KH_2PO_4, pH 7.4. Dissolve 8 g NaCl, 0.2 g KCl, 1.44 g Na_2HPO_4, and 0.24 g KH_2PO_4 in 800 mL distilled water. Adjust pH to 7.4 using hydrochloric acid or sodium hydroxide and add distilled water to 1 L. Sterilize by autoclaving and store at 4 °C.

2.2 Gel Retardation Assay

1. 10× TBE electrophoresis buffer: 1 M Tris-borate and 20 mM EDTA, pH 8.3. Weigh 121.1 g Tris base, 61.8 g boric acid, and 7.4 g EDTA in a 1 L glass beaker. Add distilled water to a volume of 900 mL. Mix the solution on a magnetic stirring plate with a stir bar until all solutes are completely dissolved, and adjust pH to 8.3. Dilute 100 mL of this 10× TBE buffer to 1 L to make 1× gel running buffer. Store up to 6 months at room temperature.
2. 0.5% agarose gel: Weigh 0.2 g SeaKem Gold agarose in a 250 mL KIMAX Erlenmeyer flask. Add 1× gel running buffer to a volume of 40 mL. Heat agarose solution in a microwave until agarose is completely dissolved (*see* **Note 1**). Cool down the agarose gel solution to about 60 °C (*see* **Note 2**). Add 4 μL

SYBR Safe DNA gel stain into agarose gel and mix gently to avoid air bubble formation. Pour agarose gel into an electrophoresis gel tray with a comb inserted. Remove any air bubbles under or between teeth of the comb. Allow the agarose gel to solidify before use.

3. A horizontal gel apparatus to run the *agarose gel electrophoresis*.
4. A ChemiDoc XRS+ Gel Imaging System (Bio-Rad, Hercules, CA, USA) which is set at an excitation wavelength of 302 nm using a trans-UV light and at an emission wavelength of 548–630 nm using the standard filter to allow green fluorescence to be captured. This imaging system contains the Quantity One 1-D analysis software v4.6.9.

2.3 Fluorescent Image Studies

1. Medium for fluorescent image studies: Prepare RPMI 1640 medium without serum. Store at 4 °C.
2. Human A549 cells or any other mammalian cell line of choice.
3. A humidified 5% CO_2 incubator set at 37 °C to house human A549 cells or mammalian cell line of choice.
4. Disposable culture plates and tissue culture dishes: 24-well culture plates, 100-mm, and 35-mm tissue culture dishes.
5. An epifluorescent or confocal microscope to capture cellular images.

3 Methods

3.1 Assessment of Noncovalent L5a/DNA Complexes

To test whether L5a peptides can interact with plasmid DNA to form noncovalent complexes, various amounts of L5a are mixed with DNA at different nitrogen (NH_3^+)/phosphate (PO_4^-) (N/P) ratios (*see* **Note 3**) followed by separation on a 0.5% agarose gel. The mobility of L5a/DNA complexes decreases as the amount of L5a increases, indicating the formation of noncovalent L5a/DNA complexes.

3.1.1 Noncovalent Binding Between L5a and Plasmid DNA

1. Mix L5a CPPs with the pEGFP-N1 plasmid at various N/P ratios in Eppendorf tubes (Table 1) (*see* **Note 4**).
2. Incubate L5a and DNA mixtures in 1.5 mL Eppendorf tubes at 37 °C for 1 h followed by gel retardation analysis.

3.1.2 Gel Retardation Assay

1. Fill the electrophoresis tank with 1× TBE gel running buffer.
2. Place the casting tray with solidified 0.5% agarose gel into the electrophoresis tank of a horizontal gel apparatus.
3. After the formation of L5a/DNA complexes, pipette L5a/DNA complexes pre-mixed with the gel loading buffer (containing glycerol to a final concentration of 6%) into the wells of the gel.

Table 1
List of all combinations between L5a peptide and the pEGFP-N1 plasmid DNA at various N/P ratios

N/P	0	3	6	9	12	15	18
L5a	0 μL (0 nmol)	0.5 μL (18 nmol)	1 μL (36 nmol)	1.5 μL (54 nmol)	2 μL (72 nmol)	2.5 μL (90 nmol)	3 μL (108 nmol)
DNA	7.4 μL (6 nmol)	7.4 μL (6 nmol)	7.4 μL (6 nmol)	7.4 μL (6 nmol)	7.4 μL (6 nmol)	7.4 μL (6 nmol)	7.4 μL (6 nmol)
H_2O	12.6 μL	12.1 μL	11.6 μL	11.1 μL	10.6 μL	10.1 μL	9.6 μL
Total	20 μL	20 μL	20 μL	20 μL	20 μL	20 μL	20 μL

4. Electrophorese the loaded agarose gel at 100 V for 30 min.
5. After electrophoresis, capture DNA gel images (Fig. 1).
6. Quantify all data from gel images by the Quantity One 1-D analysis software (*see* **Note 5**). The mobility of L5a/DNA complexes should decrease compared to DNA alone, indicating the formation of noncovalent binding between L5a and DNA (*see* **Note 6**).

3.2 Fluorescent Image Studies of L5a/DNA Complexes in Cellular Uptake

To determine the transfection ability of L5a, L5a/DNA complexes prepared at an N/P ratio of 12 (*see* **Note 7**) are incubated with cells. Fluorescent microscopy is then used to detect the *EGFP* reporter gene expression of the pEGFP-N1 plasmid delivered by L5a at the protein level.

3.2.1 Cellular Delivery of L5a/DNA Complexes

1. Adherent cell culture: Seed 1×10^5 A549 cells (or other mammalian cell lines) into 35-mm tissue culture dishes and place dishes in a humidified 5% CO_2 incubator at 37 °C overnight to allow cells to adhere to the dishes (*see* **Note 8**).
2. Mix L5a with the pEGFP-N1 plasmid at an N/P = 12 in RPMI 1640 medium without serum (*see* **Note 9**). Combine 20 μL L5a (720 nmol), 74 μL DNA (60 nmol), and 106 μL RPMI 1640 medium to make a final volume of 200 μL of complexes.
3. Incubate L5a/DNA complexes at 37 °C for 1 h.
4. Discard the overnight culture medium. Wash cells with 1 mL PBS once.
5. Add L5a/DNA complexes to the cells and incubate dishes at 37 °C for 1 h.
6. After treatment, wash cells with 1 mL PBS once and incubate dishes at 37 °C for an additional 24–48 h.

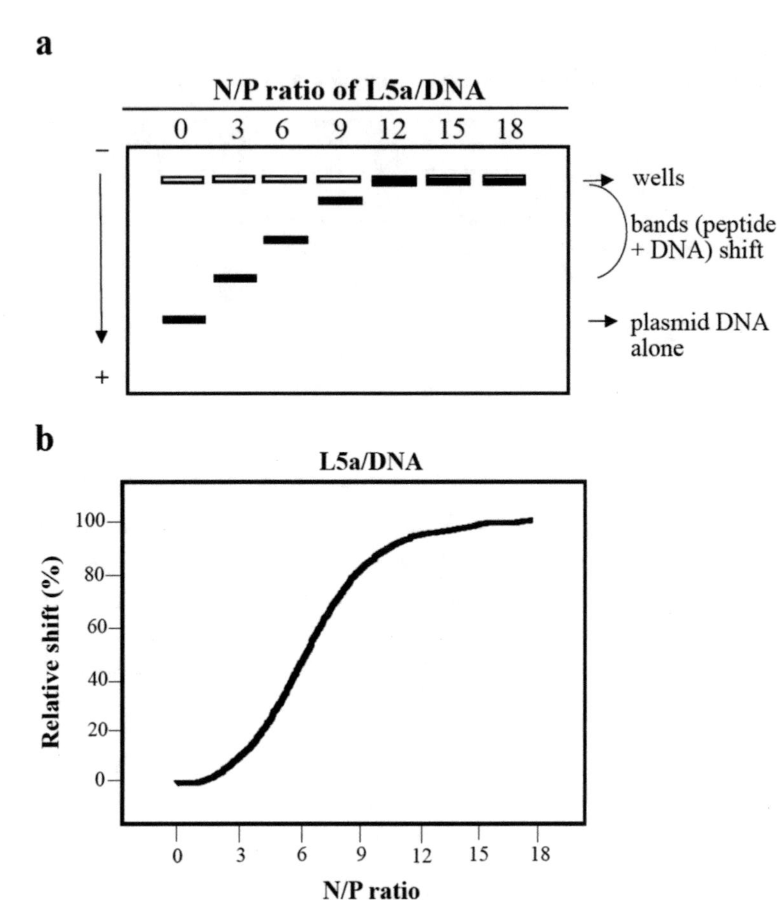

Fig. 1 Gel retardation assay for noncovalent interactions between L5a CPP and plasmid DNA at different N/P ratios. (**a**) Gel retardation assay showing the formation of CPP/DNA complexes in vitro. (**b**) The relative mobility of CPP/DNA complexes quantified from data of (**a**) panel

3.2.2 Fluorescent Microscopy

1. Detect the *EGFP* reporter gene expression using an epifluorescent or confocal microscope equipped with a digital or cooled charge-coupled device (CCD) camera.
2. Record bright-field images for cell morphology.
3. Capture green fluorescent images under GFP channel for the *EGFP* reporter gene expression in cells (*see* **Note 10**).

4 Notes

1. Agarose gel solution should be handled with caution during *microwave heating as* explosive boiling is a hazard. Agarose gel solution should be heated in a microwave oven for a few seconds/minutes and heating should be stopped right before

solution overflows. Heat-resistant gloves should be worn at all times when handling flasks containing hot agarose gel solution.

2. Hot agarose gel solution can be maintained in a 60 °C water bath. Alternatively, 40 mL of boiled agarose gel solution can be left on the benchtop to cool down for 10–20 min at room temperature.
3. Rationale of N/P ratio: The N/P ratio is a measurement of the ionic balance of noncovalent CPP/DNA complexes in vitro. It refers to the number of nitrogen residues of CPP per phosphate of DNA. Approximately, 1 μg of SR9 peptide [11] is equivalent to 7 nmol of cationic nitrogen. One μg of a 2000-bp plasmid DNA is equivalent to 3 nmol of anionic phosphate. The N/P ratio of an SR/DNA complex is $7 \div 3 = 2.3$, for example, if you mix 1 μg SR9 with 1 μg DNA together. There are two ways to obtain a desired N/P ratio. Based on the N/P fraction, you can set the denominator P constant and vary the numerator N, or vice versa. (1) Nitrogen is defined by the number of the primary cation group ($-NH_3^+$) of peptides. One μg of SR9 peptide is equivalent to 7 nmol, based on the calculation: (1×10^{-6}) g $\div$ 1423.7 g/mol $\times$ 10 positive "net" charges $\fallingdotseq 7 \times 10^{-9}$ mol = 7 nmol. On the other hand, one μg of L5a peptide is equivalent to 3.6 nmol, based on the calculation: (1×10^{-6}) g $\div$ 830.95 g/mol $\times$ 3 positive "net" charges $\fallingdotseq 3.6 \times 10^{-9}$ mol = 3.6 nmol. (2) Phosphate is defined by the number of the primary anion group ($-PO_4^-$) of nucleic acids. One μg of a 2000 bp plasmid DNA is equivalent to 3 nmol, based on the calculation: (1×10^{-6}) g $\div$ (bp in full-length $\times$ 660) g/mol $\times$ (bp in full-length $-1 + D$) $\times$ 2 strands $\fallingdotseq 3 \times 10^{-9}$ mol = 3 nmol, where D represents the digital number of primary phosphate group at the 5′ end.
4. Two μg (7.4 μL) of the pEGFP-N1 plasmid with 4733 bp in length is equivalent to 6 nmol. Ten μg (1 μL) of the L5a peptide is equivalent to 36 nmol.
5. Other software can be used to quantify the green fluorescent intensity of plasmid DNA stained by SYBR Safe, such as the UN-SCAN-IT from Silk Scientific Inc. [8] and the *public* domain ImageJ *software* (https://imagej.nih.gov/ij/) [10].
6. Rationale of gel retardation assay: Gel retardation assay also known as electrophoretic mobility shift assay (*EMSA) is an affinity electrophoresis technique commonly used to study protein–nucleic acid interactions. This method can determine whether a protein binds to a specific DNA sequence or not. In the absence of protein, a DNA labeled with radioisotope or fluorescein reveals a single band on a polyacrylamide or agarose gel during electrophoretic separation for a short period of time. This single DNA band corresponds to unbound DNA fragment and*

migrates to the longest distance on the gel. However, in the presence of a protein/peptide which can specifically bind to this DNA fragment, the protein binds to DNA and forms stable protein/DNA complexes. The migration of these larger protein/DNA complexes is retarded or shifted on the gel since they move more slowly than the unbound DNA (Fig. 1a). There will be a tendency to show the more protein concentration added to a fixed concentration of DNA probe, the more shifted protein/DNA complexes accumulate. These increasing protein/DNA complexes will ultimately *reach a plateau, once DNA is fully bound and saturated by protein (Fig. 1b).*

7. Gel retardation assay revealed that N/P ratio-dependent interactions between CPP and DNA were maximized at N/P ratios above 12 [8].
8. Cells are grown and maintained in a 100-mm tissue culture dish with 10 mL culture medium routinely. Cells can be seeded in 24-well culture plates for fluorescent image studies, and the final reaction volume is 150 μL of L5a/DNA complexes. Suspension cell culture has not been tested, but CPPs should penetrate most cells.
9. The presence of 10% FBS in media reduces the efficiency of complexing CPPs with the cargos leading to a decrease in the uptake of CPPs or their complexes/conjugates [16].
10. The efficiency of the *EGFP* reporter gene expression in the cells can be quantified by the flow cytometric analysis, or semi-quantified by the intensity of fluorescent images [8]. As stated above, the green fluorescent intensity can be converted and semi-quantified using the UN-SCAN-IT or ImageJ *software. For instance, the transfection efficiency of L5a-mediated delivery of* the *EGFP* reporter gene expression showed approximately 41-fold increase compared to the negative control in A549 cells, while *the transfection efficiency of* Lipofectamine 2000-*mediated delivery of* gene expression showed about 64-fold increase compared to the negative control [8].

Acknowledgments

This work was supported by the Center for Biomedical Research at Missouri University of Science and Technology, and the Ministry of Science and Technology, Taiwan.

References

1. Klimpel A, Lutzenburg T, Neundorf I (2019) Recent advances of anti-cancer therapies including the use of cell-penetrating peptides. Curr Opin Pharmacol 47:8–13
2. El-Andaloussi S, Holm T, Langel U (2005) Cell-penetrating peptides: mechanisms and applications. Curr Pharm Des 11:3597–3611
3. Kalafatovic D, Giralt E (2017) Cell-penetrating peptides: design strategies beyond primary structure and amphipathicity. Molecules 22: E1929
4. Heitz F, Morris MC, Divita G (2009) Twenty years of cell-penetrating peptides: from molecular mechanisms to therapeutics. Br J Pharmacol 157:195–206
5. Green M, Loewenstein PM (1988) Autonomous functional domains of chemically synthesized human immunodeficiency virus Tat transactivator protein. Cell 55:1179–1188
6. Frankel AD, Pabo CO (1988) Cellular uptake of the Tat protein from human immunodeficiency virus. Cell 55:1189–1193
7. Su R, Hu J, Zou Q, Manavalan B, Wei L (2020) Empirical comparison and analysis of web-based cell-penetrating peptide prediction tools. Brief Bioinform 21(2):408–420. https://doi.org/10.1093/bib/bby124
8. Liu BR, Huang YW, Aronstam RS, Lee HJ (2016) Identification of a short cell-penetrating peptide from bovine lactoferricin for intracellular delivery of DNA in human A549 cells. PLoS One 11:e0150439
9. Liu BR, Huang YW, Korivi M, Lo SY, Aronstam RS, Lee HJ (2017) The primary mechanism of cellular internalization for a short cell-penetrating peptide as a nano-scale delivery system. Curr Pharm Biotechnol 18:569–584
10. Lee HJ, Huang YW, Aronstam RS (2019) Intracellular delivery of nanoparticles mediated by lactoferricin cell-penetrating peptides in an endocytic pathway. J Nanosci Nanotechnol 19:613–621
11. Chang M, Huang YW, Aronstam RS, Lee HJ (2014) Cellular delivery of noncovalently-associated macromolecules by cell-penetrating peptides. Curr Pharm Biotechnol 15:267–275
12. Futaki S, Suzuki T, Ohashi W, Yagami T, Tanaka S, Ueda K, Sugiura Y (2001) Arginine-rich peptides. An abundant source of membrane-permeable peptides having potential as carriers for intracellular protein delivery J Biol Chem 276:5836–5840
13. Ulasov AV, Rosenkranz AA, Sobolev AS (2018) Transcription factors: time to deliver. J Control Release 269:24–35
14. Lee HJ, Huang YW, Chiou SH, Aronstam RS (2019) Polyhistidine facilitates direct membrane translocation of cell-penetrating peptides into cells. Sci Rep 9:9398
15. Neundorf I, Rennert R, Franke J, Kozle I, Bergmann R (2008) Detailed analysis concerning the biodistribution and metabolism of human calcitonin-derived cell-penetrating peptides. Bioconjug Chem 19:1596–1603
16. Bernkop-Schnürch A (2018) Strategies to overcome the polycation dilemma in drug delivery. Adv Drug Deliv Rev 136–137:62–72

Chapter 10

Delivery of Peptide Nucleic Acids Using an Argininocalix[4]arene as Vector

Alessia Finotti, Jessica Gasparello, Alessandro Casnati, Roberto Corradini, Roberto Gambari, and Francesco Sansone

Abstract

The importance of peptide nucleic acids (PNAs) for alteration of gene expression is nowadays firmly established. PNAs are characterized by a pseudo-peptide backbone composed of *N*-(2-aminoethyl)glycine units and have been found to be excellent candidates for antisense and antigene therapies. Recently, PNAs have been demonstrated to alter the action of microRNAs and thus can be considered very important tools for miRNA therapeutics. In fact, the pharmacological modulation of microRNA activity appears to be a very interesting approach in the development of new types of drugs. Among the limits of PNAs in applied molecular biology, the delivery to target cells and tissues is of key importance. The aim of this chapter is to describe methods for the efficient delivery of unmodified PNAs designed to target microRNAs involved in cancer, using as model system miR-221-3p and human glioma cells as in vitro experimental cellular system. The methods employed to deliver PNAs targeting miR-221-3p here presented are based on a macrocyclic multivalent tetraargininocalix[4]arene used as non-covalent vector for anti-miR-221-3p PNAs. High delivery efficiency, low cytotoxicity, maintenance of the PNA biological activity, and easy preparation makes this vector a candidate for a universal delivery system for this class of nucleic acid analogs.

Key words Peptide nucleic acids, Delivery, Cell transfection, Calixarenes, Non-viral vectors, Calixarene amphiphiles, MicroRNAs, miRNA therapeutics

Abbreviations

3′UTR	3′-Untranslated region
DCM	Dichloromethane
DMF	Dimethylformamide
DMSO	Dimethyl sulfoxide
FBS	Fetal bovine serum
Fl	Fluorescein
MiRNA/miR	MicroRNA
PBS	Phosphate-buffered saline
PNA	Peptide nucleic acid
RT-qPCR	Reverse transcription-quantitative polymerase chain reaction

Kumaran Narayanan (ed.), *Bio-Carrier Vectors: Methods and Protocols*, Methods in Molecular Biology, vol. 2211, https://doi.org/10.1007/978-1-0716-0943-9_10,

1 Introduction

1.1 MicroRNAs as Key Post-transcriptional Regulators of Gene Expression

MicroRNAs (miRNAs) are a family of small (19–25 nucleotides in length) noncoding RNAs that have been reported to finely tune gene expression following sequence-selective interaction with mRNAs targets [1]. These interactions occur at the RISC (RNA-induced silencing complex) and cause translational repression or mRNA degradation, depending on the extent of complementarity between miRNAs and mRNA target sequences [2]. While miRNAs usually interact with the 3′UTR region of mRNAs, functional binding to coding sequences and 5′UTR have also been described [1–3]. Since a single miRNA is able to interact with several mRNAs and a single mRNA may contain several signals for miRNA recognition, it is calculated that at least 20–40% of human mRNAs are targets of microRNAs [3]. With respect to effects on gene expression it is proposed that, in general, a low expression of a given miRNA is potentially associated with accumulation of the target mRNAs; conversely, a high expression of miRNAs is expected to be the cause of a low expression of the target mRNAs [4–6].

1.2 MicroRNA Therapeutics

Since the involvement of microRNAs in human pathologies is a firmly established fact, the pharmacological modulation of their activity appears to be a very interesting approach in the development of new types of drugs (miRNA therapeutics) [7–10]. For instance, miRNAs are involved in cancer; several oncomiRNAs and metastamiRNAs have been demonstrated to promote cancer cell growth and tumor invasion; among the mRNA targets of these miRNA, several tumor-suppressor mRNAs have been proposed, including PTEN, Rb1, SOCS1, $p27^{Kip1}$, PUMA, RB [11–14]. Conversely, tumor-suppressor miRNAs targeting oncoprotein-coding mRNAs have been reported [13–15]. Accordingly, these tumor-specific miRNAs can be considered molecular markers important for tumor diagnosis and prognosis, as well as important targets of therapeutic interventions [11–15].

1.3 Peptide Nucleic Acids and microRNA Therapeutics

With respect to miRNA therapeutics, peptide nucleic acid (PNA)-based molecules are appealing [16]. In PNAs, the pseudo-peptide backbone is composed of unnatural *N*-(2-aminoethyl)glycine units [17]; thus, they are resistant to both nucleases and proteases [18, 19] and, more importantly, hybridize with high affinity to complementary sequences of single-stranded RNA and DNA, forming Watson-Crick double helices [20]. For these reasons, PNAs were found to be excellent candidates for antisense and antigene therapies [21–24]. Recently, PNAs have been shown to be able of altering biological functions of microRNAs, both in vitro and in vivo [25–33]. With respect to PNA-based targeting of miRNAs involved in cancer, we have recently reported that PNAs directed against miR-221-3p and miR-155-5p inhibit tumor cell growth and induce apoptosis [34].

1.4 Delivery of PNAs

The major limit in the use of PNA for alteration of gene expression is the low uptake by eukaryotic cells [35]. In order to solve this drawback, several approaches have been considered, including the delivery of PNA analogs with liposomes and microspheres [36–41]. One of the possible strategies is to link PNAs to polyarginine (poly-R) tails, based on the observation that these cell-membrane penetrating oligopeptides are able to facilitate uptake of conjugated molecules [42, 43]. Peptide-PNA conjugates have been shown to be efficiently incorporated in cells without the need of transfecting agents [44]. Anti-miR activity was indeed observed for instance by conjugation of PNAs to poly-R tails [44–52] or by modification of the PNA backbone with cationic amino acid side chains [53, 54].

As alternative to an expensive and time-consuming chemical modification, for PNA delivery particularly convenient is the use of carriers able to interact with the cargo in a non-covalent and reversible way. This strategy would allow in principle to make available a universal system effective with all native PNA sequences that are intended to be transported into cells. In this context, it was actually already explored the delivery of PNAs and their derivatives or analogs with liposomes [38], polymer nanoparticles [55], and by co-transfection with partially complementary DNA [56].

Inorganic nanocarriers, such as nanozeolites or mesoporous silica nanoparticles (MSNPs) [57] have also been used for cellular delivery of PNAs; for MSNPs-mediated PNA delivery, an anti-miR activity was demonstrated [58]. However, the preparation of all these systems and the PNA incorporation generally require special and often time-consuming procedures. Calixarenes functionalized with guanidinium groups were shown to be suitable for the efficient delivery of nucleic acids [59–62]. In particular, a calix[4]arene in cone geometry, with amphiphilic features, functionalized with a cluster of four arginine units at the upper rim and lipophilic hexyl chains at the lower rim resulted an efficient and low toxic non-viral vector for cell transfection of DNA [61] and miRNA [62], more potent than commercial transfecting agents.

The exploited parallel arrangement of the amino acid units makes available, with respect to more classical polyarginine peptides, the primary α-amino groups that might favor the protection of the vector–nucleic acid complex from the lysosomal degradation and facilitate the release of cargo from the endosomes into the cytosol through a proton sponge effect. Very recently, the same synthetic compound was demonstrated to efficiently behave as non-covalent vector also for PNAs [63] providing an interesting tool for the delivery of these nucleic acid mimics with the perspective of boosting their therapeutic applications.

1.5 Aim of the Chapter

The aim of this chapter is to describe methods for the efficient delivery of miRNA-targeting PNAs using that amphiphilic tetra-argininocalix[4]arene as vector.

Considering that this vector is not commercially available, in the Methods section the synthetic procedures for the calixarene derivative are reported together with those relative to the production of the complexes between vector and PNA, to the transfection, the uptake analysis, the evaluation of the transfection effects and the cell viability studies study.

As a model experimental system PNA-based miR-221-3p targeting was chosen [34, 46, 48] using human glioma U251 cells [64].

2 Materials

2.1 Argininocalix[4]arene 1

For the synthesis of the vector, the commercially available materials used in the Methods section are:

1. p-Tert-butylcalix[4]arene.
2. *N,N′*-Dimethylformamide (DMF).
3. Sodium hydride (NaH, 60% in oil).
4. 1-Iodohexane.
5. Chloridric acid (HCl).
6. Dichloromethane (DCM).
7. Methanol (MeOH).
8. Sodium nitrate ($NaNO_3$).
9. Trifluoroacetic acid (TFA).
10. Magnesium sulfate ($MgSO_4$).
11. Hydrazine hydrate ($NH_2NH_2 \cdot H_2O$).
12. Pd 10% on charcoal (Pd/C).
13. N_α-Boc-N_ω-Pbf-L-Arg.
14. *N,N′*-dimethylaminopyridine (DMAP).
15. Hydroxybenzotriazole (HOBt).
16. *N*-ethyl-*N′*-(3-dimethylaminopropyl)carbodiimide hydrochloride (EDC).
17. Sodium hydrogen carbonate ($NaHCO_3$).
18. Triisopropylsilane (TIS).
19. Ethyl acetate (AcOEt).
20. Thin-layer chromatography (TLC) plates: 60 F254 silica gel and 60 RP-18 silica gel.
21. 230–400 mesh silica gel for column chromatography.

Table 1
Sequences of the PNAs employed in the described studies

PNAs	PNA sequence	Target miRNAs
Fl-PNA-a221	Fl-AEEA-AAACCCAGCAGACAATGT-NH_2	miR-221-3p
Fl-R8-PNA-a221	Fl-AEEA-R_8-AAACCCAGCAGACAATGT-NH_2	miR-221-3p
PNA-a221	H-AAACCCAGCAGACAATGT-NH_2	miR-221-3p
R8-PNA-a221	H-R_8-AAACCCAGCAGACAATGT-NH_2	miR-221-3p
R8-PNA-a221-MUT (mutations underlined)	H-R_8-AATCCCACCAGAGAAAGT-NH_2	miR-221-3p

2.2 Peptide Nucleic Acids (PNAs)

PNAs can be purchased from several companies, including Panagene Inc. (www.panagene.com). Alternatively, PNAs against miRNAs can be synthesized following the procedures described in Manicardi et al. [28]. The data here presented are based on PNAs described in the chapter by Brognara et al. [46]. The sequences of the different PNAs-a221 used are reported in Table 1. The criteria for the design of the PNAs for increasing selectivity of hybridization and low nonspecific effects are described in **Note 1**.

Figure 1 shows possible targeting of anti-apoptotic PUMA 3'UTR mRNA region by the employed miR-221-3p (a,b). Full complementarity does exist between miR-221-3p and the anti-miR PNA-a221 (sequences reported in Table 1). Some of the best characterized biological targets are also shown (c) [65–72].

2.3 Cell Culture

1. Human glioma U251 cell line [64] (Sigma-Aldrich, St.Louis, Missouri, USA).
2. RPMI 1640 medium.
3. 100 U/mL penicillin and 100 μg/mL streptomycin, final concentration.
4. Fetal bovine serum.

2.4 Cellular Uptake

1. FACScan (BD, Becton Dickinson, Franklin Lakes, New Jersey, USA).
2. 1× Dulbecco's PBS (DPBS).
3. CellQuest Pro Software (BD, Becton Dickinson, Franklin Lakes, New Jersey, USA).

2.5 RNA Extraction

1. Trypsin-EDTA.
2. FBS.
3. DPBS.
4. Tri-Reagent to lyse cells.
5. Nuclease-free water to resuspend RNA.

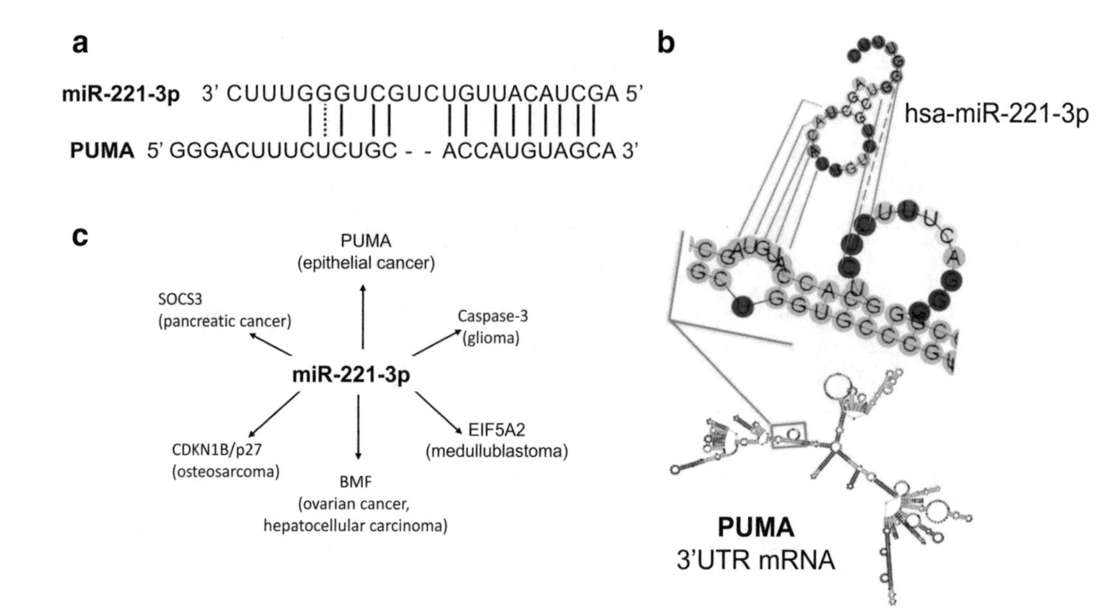

Fig. 1 Possible targeting of anti-apoptotic PUMA mRNA by miR-221-3p (**a**) and interaction between PNA-a221 and miR-221-3p (**b**). The PNA sequences discussed in this chapter are shown in Table 1. Other apoptosis-associated mRNAs described to be regulated by miR-221-3p are shown in panel c [65–72]. Modified from Brognara et al. [46]. The criteria for the design of anti-miRNA PNA molecules are discussed in **Note 1**

2.6 Real-Time Quantitative PCR of MicroRNAs

1. RT primer and miRNA quantification assay for human miR-221-3p (hsa-miR-221-3p, ID: 00524), miR-155-5p (ID: 002623), miR-210-3p (ID: 000512), and miR-96-5p (ID: 000186) (Thermo Fisher Scientific).
2. Human U6 snRNA (ID: 001973) and hsa-let-7c-5p (ID: 000379): used as references to normalize samples amount for miRNA quantification.
3. TaqMan MicroRNA Reverse Transcription Kit (Thermo Fisher Scientific).
4. CFX96 Touch Real-time PCR Detection System (Bio-Rad) to quantify miRNAs content.
5. DNA polymerase enzyme in TaqMan Universal PCR Master Mix, no AmpErase UNG 2× (Thermo Fisher Scientific).

2.7 Viability and Cell Growth

1. Muse Count and Viability Kit (Luminex Corporation, Austin, Texas, USA) to evaluate effects of argininocalix[4]arene **1** on U251 cells viability.
2. Z2 Coulter Counter (Coulter Electronics) to monitor cell growth.

2.8 Apoptosis Assay

1. Annexin V and Dead Cell assay (Luminex Corporation) to evaluate possible pro-apoptotic effects of PNA.
2. Muse Cell Analyzer instrument (Millipore Corporation) to perform Annexin V assays.

2.9 Statistics

Results are expressed as mean ± standard error of the mean (SEM). Comparisons between groups were made by using paired Student's *t* test and a one-way analysis of variance (ANOVA). Statistical significance was defined with $p < 0.01$.

3 Methods

3.1 Preparation of the Argininocalix[4] arene 1

All the following reactions (Fig. 2) are performed in N_2 atmosphere. All the procedures are performed under a fume hood.

3.1.1 Synthesis of 5,11,17,23-Tetra-tert-butyl-25,26,27,28-tetrahexyloxycalix[4] arene (2)

1. Into a two-necked round-bottom flask suspend p-tert-butylcalix[4]arene (2.5 g, 3.85 mmol) in dry DMF (40 mL), stir and cool at 0 °C with an ice bath.
2. Add NaH (60% in oil, 1.85 g, 12 equivalent) to the stirring mixture (CAUTION! Hydrogen develops at this stage).
3. After 30 min add 1-iodohexane (6.70 mL, 12 equivalent) to the reaction and, after additional 30 min, remove the ice bath leaving the reaction to proceed at room temperature for 24 h (*see* **Note 2**).

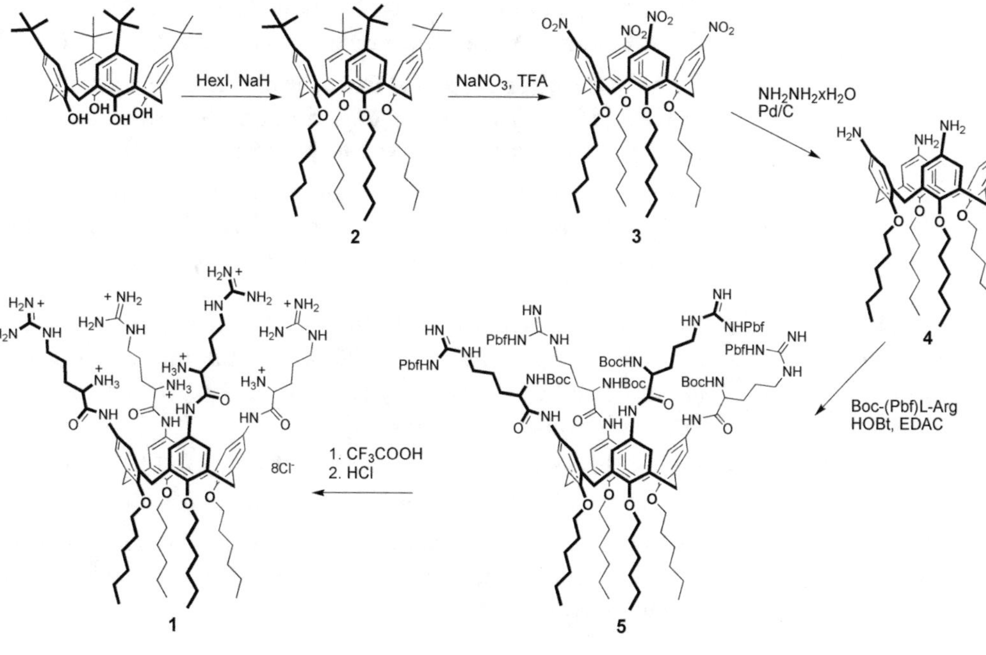

Fig. 2 Synthetic pathway for the preparation of argininocalix[4]arene **1**

4. Quench the reaction by slow addition (CAUTION! Hydrogen might develop at this stage) of aqueous 1 N HCl (50 mL) and, using a separatory funnel, extract the aqueous phase with DCM (2 × 100 mL).
5. After separation, combine the organic phases, wash then with aqueous 1N HCl (3 × 100 mL) and then evaporate DCM at the rotavapor.
6. Dissolve the residue in a flask with the minimum amount of DCM and add MeOH until a white precipitate starts to form. Leave the flask overnight in the fridge to induce the complete precipitation.
7. Filter the solid on a Buchner, wash it abundantly with MeOH, and dry it under vacuum to obtain the product as a white solid (3.05 g, 80% yield) (*see* **Note 3**).
8. Check the purity of the final compound by ESI-MS and ^{1}H NMR. It is possible to compare the collected spectra with the characterization data of the compound as reported in ref. 73.

*3.1.2 Synthesis of 5,11,17,23-Tetranitro-25,26,27,28-tetrahexyloxycalix[4]arene (**3**)*

1. Put into a two-necked round-bottom flask p-tert-butyl-tetra-hexyloxycalix[4]arene **2** (3 g, 3.04 mmol) and $NaNO_3$ (10.3 g, 120 mmol).
2. Add TFA (14 mL, 0.18 mol) dropwise upon stirring (*see* **Note 4**). The mixture becomes black and then yellow/orange. The reaction is stirred at room temperature for 24 h (*see* **Note 5**).
3. Quench the reaction by addition of water (200 mL) and extract the aqueous phase with DCM (2 × 100 mL).
4. Wash the combined organic layers with water (150 mL) and dry the organic solution over anhydrous $MgSO_4$ (*see* **Note 6**).
5. Remove the organic solvent from the filtrate at a rotavapor, add MeOH (50 mL) and accurately triturate the residue to obtain the pure product as a pale yellow powder (2.3 g, 81%).
6. Check the purity of the final compound by ESI-MS and ^{1}H NMR. It is possible to compare the collected spectra with the characterization data of the compound as reported [59].

*3.1.3 Synthesis of 5,11,17,23-Tetraamino-25,26,27,28-tetrahexyloxycalix[4]arene (**4**)*

1. Into a two-necked round-bottom flask suspend tetranitro-tetrahexyloxycalix[4]arene **3** (500 mg, 0.53 mmol) in EtOH (10 mL) and add $NH_2NH_2 \cdot H_2O$ (0.52 mL, 10.6 mmol) and Pd/C (10%) (catalytic amount).
2. Reflux the reaction mixture under stirring for 24 h equipping the central neck with a bubble condenser (*see* **Note 7**).
3. At the end of the reaction, filter the catalyst off (*see* **Note 8**), wash abundantly the retained solid and the filters with DCM (20 mL), and combine the filtrates.

4. Remove all the organic solvents under reduced pressure at a rotavapor.
5. Dissolve the residue in DCM (30 mL), dry the solution over anhydrous $MgSO_4$, filter off the solid (*see* **Note 6**), and remove from the filtrate the organic solvent at a rotavapor to obtain the pure product as a pale brown powder (415 mg, 95% yield).
6. Check the purity of the final compound by ESI-MS and ^1H NMR. It is possible to compare the collected spectra with the characterization data of the compound as reported [59].

3.1.4 Synthesis of 5,11,17,23-Tetrakis[(Nα-Boc-Nω-Pbf-L-Arg)amino]-25,26,27,28-tetra-n-hexyloxycalix[4]arene (5)

1. Stir at room temperature for 24 h in a two-necked round-bottom flask a mixture of aminocalixarene **4** (0.15 g, 0.18 mmol), N_α-Boc-N_ω-Pbf-L-Arg (0.58 g, 1.09 mmol), DMAP (0.27 mg, 2.19 mmol), HOBt (0.17 g, 1.24 mmol), and EDC (0.21 g, 1.09 mmol) in dry DCM (20 mL) (*see* **Notes 9** and **10**).
2. Quench the reaction by addition of water, separate with a separatory funnel the lower organic phase, wash it with water (2 × 25 mL) and subsequently with a saturated $NaHCO_3$ aqueous solution (2 × 25 mL).
3. Evaporate the organic solvent at the rotavapor to get a crude material that is treated by flash column chromatography (silica gel, eluent: gradient from DCM to DCM/MeOH 95:5, v/v) to isolate the pure product as a white solid (208 mg, 40% yield).
4. Check the purity of the final compound by ESI-MS and ^1H NMR. It is possible to compare the collected spectra with the characterization data of the compound as reported [61].

3.1.5 Synthesis of 5,11,17,23-Tetrakis(L-Arg-amino)-25,26,27,28-tetrakis(n-hexyloxy)calix[4]arene octa-hydrochloride (1)

1. Stir a solution of calix[4]arene **5** (80 mg, 2.8×10^{-2} mmol) in TFA/triisopropylsilane/H_2O (5 mL, 95/2.5/2.5, v/v) at room temperature for 1 h (*see* **Note 11**).
2. When the reaction is completed, remove the volatiles under reduced pressure at the rotavapor.
3. Repeatedly suspend the solid residue in AcOEt (3 × 5 mL) and remove the organic solvent under vacuum at the rotavapor (*see* **Note 12**).
4. Wash the crude solid material with distilled diethyl ether (3 × 7 mL), each time removing the supernatant liquid after sample centrifugation.
5. Dissolve the solid in a methanol solution of concentrated HCl (pH 3–4) and subsequently remove water and the volatiles at reduced pressure with the rotavapor. Repeat this operation four times (*see* **Note 13**). Compound **1** is obtained as a white solid (37 mg, 75% yield).

6. Check the purity of the final compound by ESI-MS and ^{1}H NMR. It is possible to compare the collected spectra with the characterization data of the compound as reported [61].

3.2 Production of Argininocalix[4]arene 1/PNA Complexes

1. Resuspend argininocalix[4]arene **1** in a solution of EtOH/H_2O/DMSO (2/2/1 v/v) under sterile condition.
2. Prepare a transfection mixture constituted by RPMI-1640 medium added with argininocalix[4]arene **1** at final concentration of 2.5 μM and 2 μM (final concentration) of appropriate PNA (employed PNAs are listed in Subheading 2.2).
3. Incubate the mixture for 20 min at room temperature, without serum. After the incubation, add 10% (v/v) of FBS.
4. Remove cell culture medium and replace it with the transfection mixture. Maintain the transfection mixture in contact with cells until the end of the treatment.

3.3 Human Cell Lines, Culture Conditions, and Treatment with PNA and Argininocalix[4]arene 1

1. Culture U251 cells [64] in humidified atmosphere of 5% CO_2/air in RPMI-1640 medium supplemented with 10% FBS, penicillin/streptomycin. The experiments aimed at determining the working concentrations of calix[4]arenes are discussed in **Note 14.**
2. Add to the cell culture the argininocalix[4]arene **1**/PNA-a221 complex (equivalent to 2.5 μM of argininocalix[4]arene **1** and 2 μM of PNA-a221) to evaluate PNA-a221 uptake and PNA-a221 pro-apoptotic effects. Use 2 μM R8-PNA-a221 as transfection positive control (this PNA is able to be transfected without vector employment). Monitor cell growth, according to cell number/ml, usually, after 72 h of treatment, using the Z2 Coulter Counter. The comparison of the effects of anti-miRNA PNAs with those of other anti-miRNA molecules is presented in **Note 15**.

3.4 PNAs Cellular Uptake

1. Detach cells by trypsinization, wash them twice with DPBS 1×, and suspend in 150 μL of DPBS 1×. Analyze cell suspension by FACS analysis, using FACScan, and verify cells for FITC fluorescence. For each sample acquire 30,000 events and analyze data using CellQuest Pro Software. Expected results of this methodology are briefly discussed in **Notes 16** and **17**.

3.5 RNA Extraction and Real-Time Quantitative Analysis of Free miRNAs

1. Trypsinize cells and collect them by centrifugation at 200 × *g* for 6 min at 4 °C, wash with PBS, and lyse with 1 mL of Tri-Reagent, according to manufacturer's instructions. Wash the isolated RNA once with cold 75% ethanol, dry and dissolve in an appropriate volume of nuclease-free water. Store the obtained RNA at −80 °C until the use.

2. For free microRNA quantification, perform reverse transcription using TaqMan MicroRNA Reverse Transcription Kit and miRNA-specific RT primers, following manufacturer's instructions (*see* **Note 18**). Perform real-time quantitative PCR using 30 ng of obtained cDNA for each run. Prepare an RT-qPCR mix containing cDNA, miRNA-specific PCR assay 20× and TaqMan Universal PCR Master Mix, no AmpErase UNG 2×. Run all RT-qPCR reactions, including no-template controls and RT-minus controls, in duplicate, using the CFX96 Touch Real-Time PCR Detection System. Calculate miR-221-3p relative expression using the comparative cycle threshold method and use U6 snRNA (TM:001973) and let-7c-5p (TM: 000379) as reference since it remains constant in the assayed samples by miR-profiling and quantitative RT-PCR analysis, as previously reported [34, 46, 48]. A comprehensive discussion focusing on the specificity of PNA-mediated miRNA inhibition is presented in **Note 19**. Additional information on required controls for validation of the PNA-mediated alteration of biological functions is presented in **Notes 20** and **21**.

3.6 Studies on Cell Viability

1. Evaluate cell viability after 72 h contact with argininocalix[4]arene **1** using Muse Count and Viability Kit and Muse Cell Analyzer. Perform analysis, using 50 μL of suspension cells (obtained after U251 trypsinization) added with 225 μL of Muse Count and Viability Reagent. Incubate the mixture at room temperature for 5 min, protected from the light and then acquire 1×10^3 events using Muse Cell Analyzer. Evaluate the effects of argininocalix[4]arene **1** on cell growth determining the cell number/mL using a Z2 Coulter Counter. Examples focusing on the possible cytotoxicity of argininocalix[4]arene **1** are presented in **Note 22.**

4 Notes

1. PNAs are usually designed according to the following criteria: (a) length of at least 18 bases; 18 bases ensures high stability of the PNA:RNA duplex, and suitable efficiency of the synthesis on large scale (this issue is particularly important if ex vivo and in vivo studies are programmed, requiring large quantity of PNAs), whereas longer sequences might prove difficult to obtain with proper yields and purities; (b) lack of self-complementarity both in the antiparallel and parallel orientation since the PNA:PNA duplex is more stable than PNA:DNA and PNA:RNA complexes; (c) minimal length of complementary sequences when the complete set of the transcriptome is considered, as evaluated by bioinformatic analyses (such as a BLAST search); after this estimation, it is also important to

store the information of possible interfering sequences for the interpretation of biological data; we normally consider off targets with almost complete complementarity (i.e., 16–17 bases/18) as severe interferent and redesign is needed if these are biologically relevant; (d) when alternative sequences are suitable, targeting of the "seed region" (an essential element for miRNA function) [50] is preferred.

2. To verify the completion of the reaction, carry out a TLC analysis (eluent: DCM/hexane 3/1, v/v) on a sample. The spots of the products and intermediates are detected by irradiation of the TLC plate with a UV lamp. The desired compound has an rf = 0.8, while possible intermediate products due to partial alkylation and containing free phenolic OH groups show lower rf and can be revealed by treatment of the TLC plate with a saturated aqueous solution of $FeCl_3$ followed by gentle heating (brown coloration of the spot).
3. If the purification by recrystallization is not sufficient to remove the persistent impurities and by-products (detected by TLC, *see* **Note 2**), a flash chromatography column on silica gel using hexane as eluent represents a convenient methods to obtain pure p-*tert*-butyl-tetrahexyloxycalix[4]arene **2**.
4. The reaction can conveniently be performed also in a mixture DCM/TFA 5/7 (v/v) dissolving the calixarene in DCM, with comparable results in terms of yields.
5. To verify the completion of the reaction, take a sample, and check it by TLC (eluent: DCM). Due to the heterogenicity of the reaction, be sure to sample also a portion of solid together with an aliquot of solution. The reaction is complete when the spot corresponding to the reagent **2** is disappeared and only a spot with rf = 0.30 corresponding to the product is present. The spots on the TLC plate are visible under UV lamp.
6. The hydrated $MgSO_4$ should be filtered off on a pleated paper filter and abundantly washed with DCM.
7. To verify the completion of the reaction, this is monitored by TLC (eluent: DCM) to check the disappearance of the starting calixarene (rf = 0.30). Due to the reduction of the nitro groups to amines, the product, as the partially reduced intermediates, are retained at the TLC starting line. The TLC plate, after elution, can be treated with an acidic solution of ninhydrin in EtOH, followed by heating, to check the presence at the starting line of a spot that becomes purple/red corresponding to amino containing calixarenes. Due to the difficulty in separating on TLC plate the partially reduced intermediates (calixarenes still bearing residual nitro groups) from the completely reduced product, it could be convenient to monitor the

evolution of the reaction by electrospray mass spectrometry. This technique can easily evidence the presence of possible intermediates still bearing nitro groups.

8. The Pd/C catalyst is a very thin solid. At least two superimposed paper filters are necessary to efficiently filter it off from the reaction mixture. However, the catalyst could not be completely retained and the filtrate could still present a pale black/gray color. In this latter case, the filtrate must be filtered again using the same procedure.
9. Alternatively, the reaction can be conveniently carried out in a microwave reactor performing two cycles (1 h each) at 200 W and 25 °C.
10. The completion of the reaction is monitored by TLC (eluent: DCM/MeOH 96/4, v/v) and spots revealed under the UV lamp. The product has an rf = 0.32.
11. The progression of the reaction is monitored by TLC (eluent: DCM/MeOH 96/4, v/v) to verify the disappearance of the starting fully protected calixarene (rf = 0.32) or, even better, by ESI–Mass Spectrometry. In fact, the removal of even a single protecting group is sufficient to retain the corresponding product at the starting line in the TLC plate. An ESI-MS analysis in positive ionization mode is then necessary to verify the possible presence of intermediates due to partial deprotection in addition to the final fully deprotected compound.
12. The repetition of this operation helps the complete removal of residual TFA.
13. With this procedure, the trifluoroacetate anions of the resulting TFA octa-salt are replaced by chloride anions.
14. In order to determine the concentrations of argininocalix[4] arene **1** to be employed for in vitro studies on target cells, the IC_{50} after treatment for different days should be always determined. In order to avoid to use antiproliferative (and possibly cytotoxic) concentrations, the argininocalix[4]arene **(1)** should be used at concentration lower than the IC_{50} values. In any case, experiments similar to those depicted in Fig. 3 might be considered [62, 63].
15. Unlike commercially available anti-miRNAs, which needs continuous administrations, a single administration of anti-miRNA PNAs complexed to argininocalix[4]arene **1** is sufficient to obtain the biological effects on microRNA activity [47]. In order to have a PNA to be used as positive control, PNAs carrying an octaarginine (R8) peptide conjugated at N-terminus of the PNA chain should be considered. We have reported that the delivery of these R8-carrying PNAs approaches 100% (i.e., uptake in 100% of the target cell

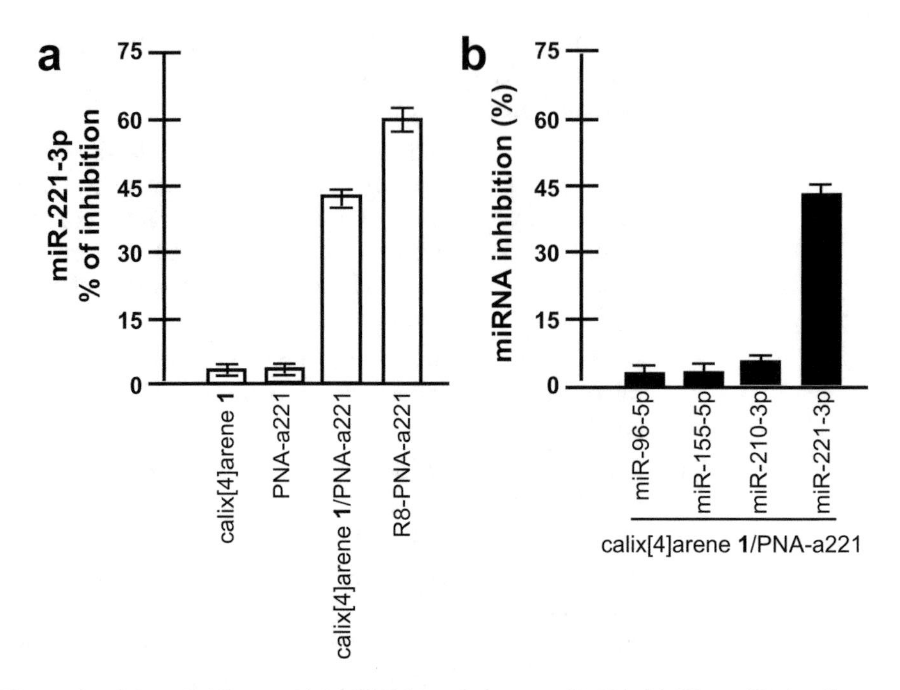

Fig. 3 Effects of argininocalix[4]arene **1**/a221PNA formulation on miR-221. (**a**) Glioma U251 cells were treated with argininocalix[4]arene **1,** PNA-a221, argininocalix[4]arene **1**/PNA-a221 formulation and R8-PNA-a221, as indicated. After 48 h, RNA was isolated and the hybridization to an miR-221-3p probe determined by RT-qPCR. Argininocalix[4]arene 1 was used at 2.5 μM; PNAs were used at 2 μM. (**b**) Lack of inhibitory effects on miR-96-5p, miR-155-5p, and miR-210-3p after treatment of U251 cells with the argininocalix[4]arene 1/PNA-a221 formulation. *See* Table 1 for PNA sequences. Modified from Gasparello et al. [63]

population) [46]; this R8-PNA conjugation is easily realized during PNA solid-phase synthesis using the same reagents and solvents.

16. When U251 cells are cultured in the presence of PNA-a221 and R8-PNA-a221 (*see* **Note 1** for the design of miRNA-interfering PNAs), a clear-cut result should be obtained as depicted in Fig. 4. First of all, culturing of U251 cells with the PNA-a221 (lacking the R8 peptide) is not associated with uptake by the cells (blue line). On the contrary, high level uptake by R8-PNA-a221 should be reproducibly obtained (orange line). Interestingly, incubation of U251 cells with the argininocalix[4]arene **1**/PNA-a221 formulation will lead to uptake of the Fl-label PNA-a221 (green line). (*See* **Note 17** for further considerations on the efficiency of PNA uptake by target cells.)
17. In the case, FACS analyses are programmed to demonstrate efficient uptake of the anti-miRNA PNAs, fluorescein-conjugated PNAs might be very useful to be complexed to the argininocalix[4]arene (**1**) compound. In any case, the FACS results should be confirmed by microscope-assisted

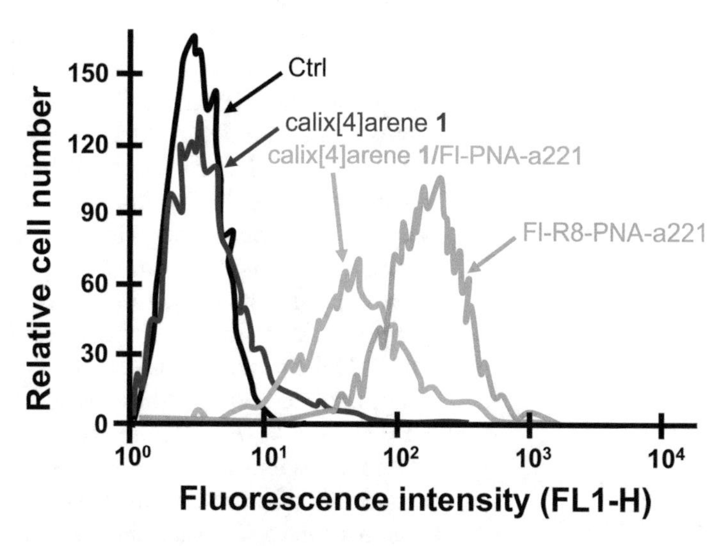

Fig. 4 Cellular uptake of PNAs. FACS analysis showing the uptake by glioma U251 cells of fluorescein-labeled a221PNAs. U251 cells were untreated (Ctrl) or cultured for 48 h with the Fl-PNA-a221 (blue line), argininocalix[4]arene **1**/Fl-PNA-a221 formulation (green line) or Fl-R8-PNA-a221 (orange line), as indicated. Modified from Gasparello et al. [63]

analyses to demonstrated that PNAs are internalized within target cells. Confocal analyses might in addition be very useful if PNA distribution within intracellular compartments should be determined, as pointed out in Brognara et al. [46].

18. It should be emphasized that the binding of PNA to microRNA does not alter the production of miRNA, and that there is no evidence so far that PNA can induce degradation of the target microRNA. Thus, the measurements performed using RT-qPCR do not measure the microRNA expression, but rather indicate its *bioavailability*, which is dependent on the amount of free microRNA which is not sequestered by PNA. For a discussion of the cellular fate of microRNA after PNA/LNA binding *see* Torres et al. [74].

19. As for specificity on PNA-mediated miRNA inhibition (verified quantifying free miRNAs using RT-qPCR), the results obtained demonstrate that first of all culturing U251 cells in the presence of the PNA-a221 (lacking the R8 peptide) is not associated with miR-221-3p inhibition (Fig. 3a). On the contrary, miR-221-3p hybridization signals are deeply inhibited when the U251 cells have been cultured in the presence of R8-PNA-a221. Interestingly, the incubation of U251 cells with the argininocalix[4]arene **1**/PNA-a221 formulation leads to results very similar to that obtained with the R8-PNA-a221 and fully compatible with the concept that high level of cellular uptake is obtained by the delivery of the

PNA-a221 with the argininocalix[4]arene **1** (Fig. 3a). To further confirm that the effect is sequence-specific the effects of argininocalix[4]arene **1** delivered PNA-a221 on other miRNAs should be determined. The expected result is that hybridization specific for other miRNAs expressed in U251 cells (for instance, miR-96-5p, miR-155-5p, and miR-210-3p) was found unchanged following delivery of the PNA-a221 with the argininocalix[4]arene **1** (Fig. 3b). In fact, the representative experiments shown in Figure 3b demonstrate no effect of argininocalix[4]arene **1** delivered PNA-a221 on miR-96-5p, miR-155-5p, and miR-210-3p. Altogether these experiments support the concept that the effects of argininocalix[4]arene **1** delivered PNA-a221 are sequence-specific (*see* **Note 20** for a further discussion on the effects of PNAs on the hybridization efficiency of targeted miRNA). Of course, the effects of PNA against microRNAs should also be analyzed on the expression of miRNA-regulated functions and/or target mRNAs.

20. Explanation of the effects of PNAs delivered by the argininocalix[4]arene **1** on microRNA accumulation. PNAs might interact very stably with target mRNAs. Therefore, before performing RT-qPCR, it is necessary to demonstrate that PNAs are not co-purified with target mRNAs. High-quality RNA preparation is necessary. Complementary analyses (for instance, Northern blotting) might also be considered. Studies suggesting a PNA-mediated effect on microRNA content should be in any case accompanied by studies on the effects on the expression of mRNAs demonstrated to be target of the studied microRNAs [46].
21. The effects of PNA against microRNAs should also be analyzed on the expression of miRNA-regulated functions. In our example, several biological functions have been firmly established to be target molecules of miR-221-3p, such as apoptosis [65–72]. In the context of the development of anti-tumor approaches, the pro-apoptotic activity following delivery of PNA-a221 by argininocalix[4]arene **1** to U251 cells is of great interest. This is described in Fig. 5. First of all, apoptosis is not induced by a mutated R8-PNA-a221 (Fig. 5a) or by the PNA-a221 lacking the R8 peptide (Fig. 5b). By contrast, high levels of apoptosis were obtained when 1/PNA-a221 formulation is employed (Fig. 5b–d).
22. The results shown in Fig. 6 were performed in order to verify whether argininocalix[4]arene **1** was to some extent cytotoxic. Figure 6a, b shows that argininocalix[4]arene **1** employed at concentrations equal or lower than 5 μM did not reduce the extent of viable cells (panel a) and did not exhibit antiproliferative effects (panel b). Moreover, argininocalix[4]arene **1** did not induce alteration of morphology of U251 cells.

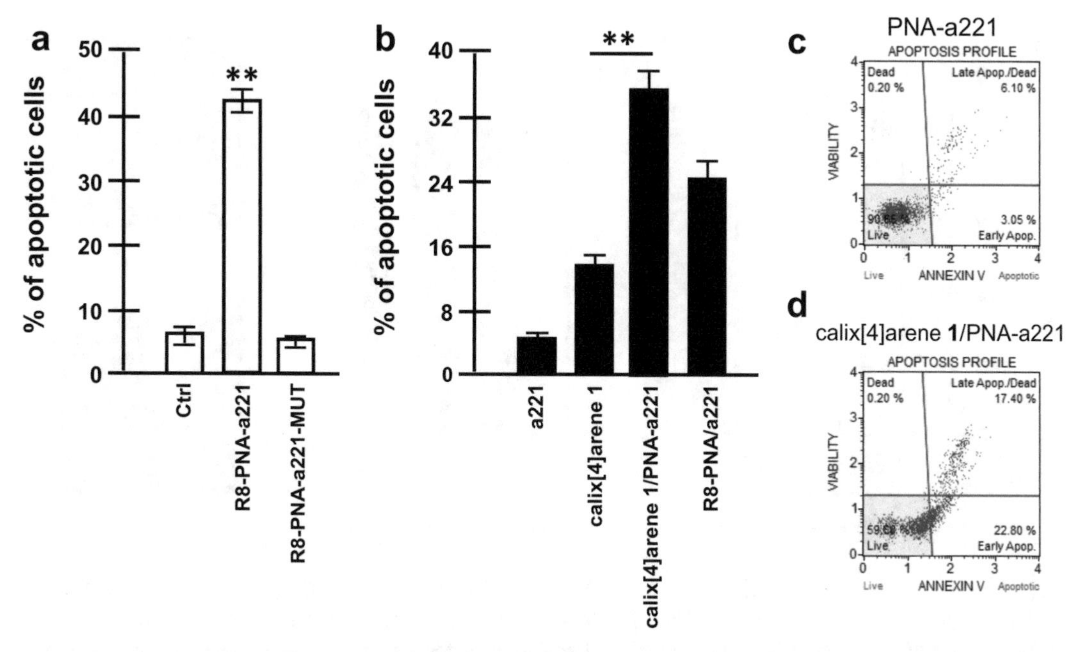

Fig. 5 (**a**) Induction of apoptosis of U251 cells by R8-PNA-a221, but not by the mutated R8-PNA-a221-MUT (*see* Table 1 for PNA sequences). (**b**) Effects of argininocalix[4]arene **1**/PNA-a221 formulation on U251 apoptosis. U251 cells were cultured with argininocalix[4]arene **1**, PNA-a221, argininocalix[4]arene **1**/PNA-a221 formulation, and R8-PNA-a221 as indicated and the percentage of apoptotic cells was determined. After 72 h, cells were detached and Annexin V assay was performed. Percentage of apoptotic cells in graphs refers to the sum of early apoptotic and late apoptotic cells. (**c**, **d**) Representative apoptosis profile of U251 cells treated with PNA-a221 (**c**) or argininocalix[4]arene **1**/PNA-a221 formulation (**d**). Modified from Brognara et al. [46] and Gasparello et al. [63]

Acknowledgments

This work was supported by Consorzio Interuniversitario di Biotecnologie, Ministero dell'Istruzione, dell'Università e della Ricerca [PRIN-2017 project 2017E44A9P, bacHound; PRIN-2009 project 20093N774P, Riconoscimento molecolare di micro-RNA (miR) mediante PNA modificati: dalla struttura alla attività; COMP-HUB initiative, Departments of Excellence Program 2018-2022], Associazione Italiana per la Ricerca sul Cancro [project IG 13575, Peptide nucleic acids targeting oncomiR and tumor-suppressor miRNAs: cancer diagnosis and therapy], EU-FP7 [THALAMOSS Project—Thalassemia Modular Stratification System for Personalized Therapy of Β-Thalassemia; n.306201-FP7-HEALTH-2012-INNOVATION-1], Wellcome Trust, AIFA, and by Fondazione Fibrosi Cistica [Project "Revealing the microRNAs-transcription factors network in cystic fibrosis: from microRNA therapeutics to precision medicine (CF-miRNA-THER)," FFC#7/2018].

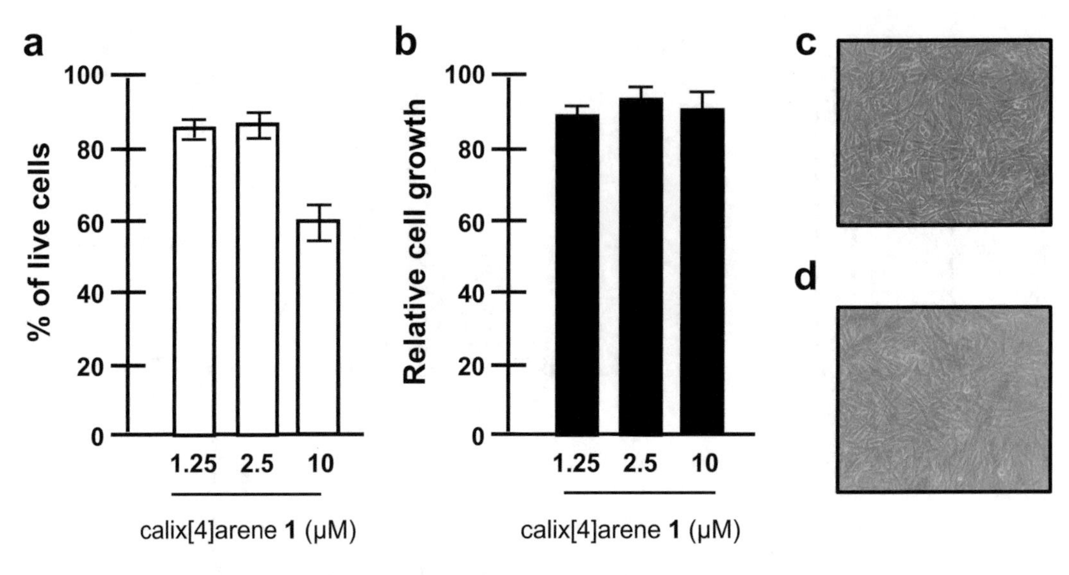

Fig. 6 Effects of argininocalix[4]arene **1** on viability, cell proliferation, and morphology of U251 cells. (**a**) Viability profile in U251 cells was reported for three incremental concentrations of argininocalix[4]arene **1**: 1.25, 2.5, and 10 μM. The assay was conducted after 72-h transfection. (**b–d**) Lack of alteration of cell growth and morphology of glioma U251 cells treated for 72 h with the indicated concentrations of argininocalix[4] arene **1** (**b**). In panel (**c**) and (**d**), representative pictures are shown of control untreated U251 cells (**c**) and U251 treated with 2.5 μM of argininocalix[4]arene **1** (**d**). Modified from Gasparello et al. [63]

References

1. He L, Hannon GJ (2010) MicroRNAs: small RNAs with a big role in gene regulation. Nat Rev Genet 5:522–531
2. Griffiths-Jones S (2004) The microRNA registry. Nucleic Acids Res 32:D109–D111
3. Monga I, Kumar M (2019) Computational resources for prediction and analysis of functional miRNA and their targetome. Methods Mol Biol 1912:215–250
4. Lim LP, Lau NC, Garrett-Engele P et al (2005) Microarray analysis shows that some microRNAs downregulate large numbers of target mRNAs. Nature 433:769–773
5. Filipowicz W, Jaskiewicz L, Kolb FA et al (2005) Post-transcriptional gene silencing by siRNAs and miRNAs. Curr Opin Struct Biol 15:331–341
6. Saliminejad K, Khorram Khorshid HR, Soleymani Fard S et al (2019) An overview of microRNAs: biology, functions, therapeutics, and analysis methods. J Cell Physiol 234:5451–5465
7. Laina A, Gatsiou A, Georgiopoulos G et al (2018) RNA therapeutics in cardiovascular precision medicine. Front Physiol 9:953
8. Bardin P, Sonneville F, Corvol H et al (2018) Emerging microRNA therapeutic approaches for cystic fibrosis. Front Pharmacol 9:1113
9. Finotti A, Fabbri E, Lampronti I et al (2019) MicroRNAs and long non-coding RNAs in genetic diseases. Mol Diagn Ther 23:155–171
10. Lima JF, Cerqueira L, Figueiredo C et al (2018) Anti-miRNA oligonucleotides: a comprehensive guide for design. RNA Biol 15:338–352
11. Kwok GT, Zhao JT, Weiss J et al (2017) Translational applications of microRNAs in cancer, and therapeutic implications. Noncoding RNA Res 2:143–150
12. Li Y, Wang Y, Shen X, Han X (2019) miR-128 functions as an oncomiR for the downregulation of HIC1 in breast cancer. Front Pharmacol 10:1202
13. Kim J, Siverly AN, Chen D et al (2016) Ablation of miR-10b suppresses oncogene-induced mammary tumorigenesis and metastasis and reactivates tumor-suppressive pathways. Cancer Res 76:6424–6435
14. Gambari R, Brognara E, Spandidos DA et al (2016) Targeting oncomiRNAs and mimicking

tumor suppressor miRNAs: new trends in the development of miRNA therapeutic strategies in oncology (Review). Int J Oncol 49:5–32

15. Elhefnawi M, Salah Z, Soliman B (2019) The promise of miRNA replacement therapy for hepatocellular carcinoma. Curr Gene Ther 9:290–304
16. Gambari R (2014) Peptide nucleic acids: a review on recent patents and technology transfer. Expert Opin Ther Pat 24:267–294
17. Nielsen PE, Egholm M, Berg RH et al (1991) Sequence-selective recognition of DNA by strand displacement with a thymine-substituted polyamide. Science 254:1497–1500
18. Demidov VV, Potaman VN, Frank-Kamenetskii MD et al (1994) Stability of peptide nucleic acids in human serum and cellular extracts. Biochem Pharmacol 48:1310–1313
19. Nielsen PE (2001) Targeting double stranded DNA with peptide nucleic acid (PNA). Curr Med Chem 8:545–550
20. Egholm M, Buchardt O, Christensen L et al (1993) PNA hybridizes to complementary oligonucleotides obeying the Watson-Crick hydrogen-bonding rules. Nature 365:566–568
21. Nielsen PE (2010) Gene targeting and expression modulation by peptide nucleic acids (PNA). Curr Pharm Des 16:3118–3123
22. Tonelli R, McIntyre A, Camerin C et al. (2012) Antitumor activity of sustained N-myc reduction in rhabdomyosarcomas and transcriptional block by antigene therapy. Clin Cancer Res 18:796–807
23. Gambari R (2001) Peptide nucleic acids (PNAs): a tool for the development of gene expression modifiers. Curr Pharm Des 7:1839–1862
24. Montagner G, Bezzerri V, Cabrini G et al (2017) An antisense peptide nucleic acid against *Pseudomonas aeruginosa* inhibiting bacterial-induced inflammatory responses in the cystic fibrosis IB3-1 cellular model system. Int J Biol Macromol 99:492–498
25. Cheng CJ, Bahal R, Babar IA et al (2015) MicroRNA silencing for cancer therapy targeted to the tumour microenvironment. Nature 518:107–110
26. Gupta A, Quijano E, Liu Y et al (2017) Antitumor activity of miniPEG-γ-modified PNAs to inhibit microRNA-210 for cancer therapy. Mol Ther Nucleic Acids 9:111–119
27. Quijano E, Bahal R, Ricciardi A et al (2017) Therapeutic peptide nucleic acids: principles, limitations, and opportunities. Yale J Biol Med 90:583–598
28. Manicardi A, Gambari R, de Cola L et al (2018) Preparation of anti-miR PNAs for drug development and nanomedicine. Methods Mol Biol 1811:49–63
29. Fabbri E, Tamanini A, Jakova T et al (2017) A peptide nucleic acid against microRNA miR-145-5p enhances the expression of the cystic fibrosis transmembrane conductance regulator (CFTR) in Calu-3 cells. Molecules 23:71
30. Mercurio S, Cauteruccio S, Manenti R et al (2019) miR-7 knockdown by peptide nucleic acids in the ascidian Ciona intestinalis. Int J Mol Sci 20:5127
31. Seo YE, Suh HW, Bahal R et al (2019) Nanoparticle-mediated intratumoral inhibition of miR-21 for improved survival in glioblastoma. Biomaterials 201:87–98
32. Sajadimajd S, Yazdanparast R, Akram S (2016) Involvement of Numb-mediated HIF-1α inhibition in anti-proliferative effect of PNA-antimiR-182 in trastuzumab-sensitive and -resistant SKBR3 cells. Tumour Biol 37:5413–5426
33. Amato F, Tomaiuolo R, Nici F et al (2014) Exploitation of a very small peptide nucleic acid as a new inhibitor of miR-509-3p involved in the regulation of cystic fibrosis disease-gene expression. Biomed Res Int 2014:610718
34. Milani R, Brognara E, Fabbri E et al (2019) Targeting miR-155-5p and miR-221-3p by peptide nucleic acids induces caspase-3 activation and apoptosis in temozolomide-resistant T98G glioma cells. Int J Oncol 55:59–68
35. Rasmussen FW, Bendifallah N, Zachar V et al (2006) Evaluation of transfection protocols for unmodified and modified peptide nucleic acid (PNA) oligomers. Oligonucleotides 16:43–57
36. Cortesi R, Mischiati C, Borgatti M et al (2004) Formulations for natural and peptide nucleic acids based on cationic polymeric submicron particles. AAPS PharmSci 6:10–21
37. Nastruzzi C, Cortesi R, Esposito E et al (2000) Liposomes as carriers for DNA-PNA hybrids. J Control Release 68:237–249
38. Ringhieri P, Avitabile C, Saviano M et al (2016) The influence of liposomal formulation on the incorporation and retention of PNA oligomers. Colloids Surf B Biointerfaces 145:462–469
39. Chiarantini L, Cerasi A, Millo E et al (2005) Antisense peptide nucleic acid delivered by core-shell microspheres. J Control Release 101:397–398
40. Mischiati C, Sereni A, Finotti A et al (2004) Complexation to cationic microspheres of double-stranded peptide nucleic acid-DNA

chimeras exhibiting decoy activity. J Biomed Sci 11:697–704
41. Chiarantini L, Cerasi A, Millo E et al (2006) Enhanced antisense effect of modified PNAs delivered through functional PMMA microspheres. Int J Pharm 324:83–91
42. Albertshofer K, Siwkowski AM, Wancewicz EV et al (2005) Structure-activity relationship study on a simple cationic peptide motif for cellular delivery of antisense peptide nucleic acid. J Med Chem 48:6741–6749
43. Kaihatsu K, Huffman KE, Corey DR (2004) Intracellular uptake and inhibition of gene expression by PNAs and PNA-peptide conjugates. Biochemistry 43:14340–14347
44. Fabbri E, Manicardi A, Tedeschi T et al (2011) Modulation of the biological activity of microRNA-210 with peptide nucleic acids (PNAs). ChemMedChem 6:2192–2202
45. Torres AG, Threlfall RN, Gait MJ (2011) Potent and sustained cellular inhibition of miR-122 by lysine-derivatized peptide nucleic acids (PNA) and phosphorothioate locked nucleic acid (LNA)/2′-O-methyl (OMe) mixmer anti-miRs in the absence of transfection agents. Artif DNA 3:71–78
46. Brognara E, Fabbri E, Bazzoli E et al (2014) Uptake by human glioma cell lines and biological effects of a peptide-nucleic acids targeting miR-221. J Neuro-Oncol 118:19–28
47. Brognara E, Fabbri E, Bianchi N et al (2014) Molecular methods for validation of the biological activity of peptide nucleic acids targeting microRNAs. Methods Mol Biol 1095:165–176
48. Brognara E, Fabbri E, Montagner G et al (2016) High levels of apoptosis are induced in human glioma cell lines by co-administration of peptide nucleic acids targeting miR-221 and miR-222. Int J Oncol 48:1029–1038
49. Shiraishi T, Nielsen PE (2006) Enhanced delivery of cell-penetrating peptide-peptide nucleic acid conjugates by endosomal disruption. Nat Protoc 1:633–636
50. Ghavami M, Shiraishi T, Nielsen PE (2019) Cooperative cellular uptake and activity of octaarginine antisense peptide nucleic acid (PNA) conjugates. Biomol Ther 9:554
51. Soudah T, Khawaled S, Aqeilan RI et al (2019) AntimiR-155 cyclic peptide-PNA conjugate: synthesis, cellular uptake, and biological activity. ACS Omega 4:13954–13961
52. Brognara E, Fabbri E, Aimi F et al (2012) Peptide nucleic acids targeting miR-221 modulate p27Kip1 expression in breast cancer MDA-MB-231 cells. Int J Oncol 41:2119–2127
53. Manicardi A, Fabbri E, Tedeschi T et al (2012) Cellular uptakes, biostabilities and anti-miR-210 activities of chiral arginine-PNAs in leukaemic K562 cells. Chembiochem 13:1327–1337
54. Verona MD, Verdolino V, Palazzesi F et al (2017) Focus on pna flexibility and rna binding using molecular dynamics and metadynamics. Sci Rep 7:42799
55. Liu C, Wang J, Huang S et al (2018) Self-assembled nanoparticles for cellular delivery of peptide nucleic acid using amphiphilic N,N,N-trimethyl-O-alkyl chitosan derivatives. J Mater Sci Mater Med 29:114
56. Borgatti M, Breda L, Cortesi R et al (2002) Cationic liposomes as delivery systems for double-stranded PNA-DNA chimeras exhibiting decoy activity against NF-kappaB transcription factors. Biochem Pharmacol 64:609–616
57. Beavers KR, Werfel TA, Shen T et al (2016) Porous silicon and polymer nanocomposites for delivery of peptide nucleic acids as anti-microRNA therapies. Adv Mater 28:7984–7992
58. Bertucci A, Prasetyanto EA, Septiadi D et al (2015) Combined delivery of temozolomide and anti-miR221 PNA using mesoporous silica nanoparticles induces apoptosis in resistant glioma cells. Small 11:5687–5695
59. Sansone F, Dudič M, Donofrio G et al (2006) DNA condensation and cell transfection properties of guanidinium calixarenes: dependence on macrocycle lipophilicity, size, and conformation. J Am Chem Soc 128:14528–14536
60. Bagnacani V, Franceschi V, Fantuzzi L et al (2012) Lower rim guanidinocalix[4]arenes: macrocyclic nonviral vectors for cell transfection. Bioconjug Chem 23:993–1002
61. Bagnacani V, Franceschi V, Bassi M et al (2013) Arginine clustering on calix[4]arene macrocycles for improved cell penetration and DNA delivery. Nat Commun 4:1721
62. Gasparello J, Lomazzi M, Papi C et al (2019) Efficient delivery of microRNA (miRNA) and anti-miRNA molecules using an argininocalix [4]arene macrocycle. Mol Ther Nucleic Acids 18:748–763
63. Gasparello J, Manicardi A, Casnati A et al (2019) Efficient cell penetration and delivery of peptide nucleic acids by an argininocalix[4] arene. Sci Rep 9:3036
64. Cao X, Gu Y, Jiang L et al (2013) A new approach to screening cancer stem cells from the U251 human glioma cell line based on cell growth state. Oncol Rep 29:1013–1018
65. Zhang C, Zhang J, Zhang A et al (2010) PUMA is a novel target of miR-221/222 in

human epithelial cancers. Int J Oncol 37:1621–1626

66. Xie J, Wen JT, Xue XJ et al (2018) MiR-221 inhibits proliferation of pancreatic cancer cells via down regulation of SOCS3. Eur Rev Med Pharmacol Sci 22:1914–1921
67. Zhou QY, Peng PL, Xu YH (2019) MiR-221 affects proliferation and apoptosis of gastric cancer cells through targeting SOCS3. Eur Rev Med Pharmacol Sci 23:9427–9435
68. Ergun S, Arman K, Temiz E et al (2014) Expression patterns of miR-221 and its target Caspase-3 in different cancer cell lines. Mol Biol Rep 41:5877–5881
69. Hu XH, Zhao ZX, Dai J et al (2019) MicroRNA-221 regulates osteosarcoma cell proliferation, apoptosis, migration, and invasion by targeting CDKN1B/p27. J Cell Biochem 120:4665–4674
70. Xie X, Huang Y, Chen L et al (2018) miR-221 regulates proliferation and apoptosis of ovarian cancer cells by targeting BMF. Oncol Lett 16:6697–6704
71. Gramantieri L, Fornari F, Ferracin M et al (2009) MicroRNA-221 targets Bmf in hepatocellular carcinoma and correlates with tumor multifocality. Clin Cancer Res 15:5073–5081
72. Yang Y, Cui H, Wang X (2019) Downregulation of EIF5A2 by miR-221-3p inhibits cell proliferation, promotes cell cycle arrest and apoptosis in medulloblastoma cells. Biosci Biotechnol Biochem 83:400–408
73. Kenis PJA, Noordman OFJ, Schonherr H et al (1998) Supramolecular materials: molecular packing of tetranitrotetrapropoxycalix[4]arene in highly stable films with second-order nonlinear optical properties. Chem Eur J 4:1225–1234
74. Torres AG, Fabani MM, Vigorito E, Gait MJ (2011) MicroRNA fate upon targeting with anti-miRNA oligonucleotides as revealed by an improved Northern-blot-based method for miRNA detection. RNA 17:933–943

Part IV

Hybrid Carriers and Other Bio Agents

Chapter 11

Cell-Derived Nanovesicles as Exosome-Mimetics for Drug Delivery Purposes: Uses and Recommendations

Yi-Hsuan Ou, Shui Zou, Wei Jiang Goh, Jiong-Wei Wang, Matthias Wacker, Bertrand Czarny, and Giorgia Pastorin

Abstract

Cell-derived Drug Delivery Systems (DDSs), particularly exosomes, have grown in popularity and have been increasingly explored as novel DDSs, due to their intrinsic targeting capabilities. However, clinical translation of exosomes is impeded by the tedious isolation procedures and poor yield. Cell-derived nanovesicles (CDNs) have recently been produced and proposed as exosome-mimetics. Various methods for producing exosome-mimetics have been developed. In this chapter, we present a simple, efficient, and cost-effective CDNs production method that uses common laboratory equipment (microcentrifuge) and spin cups. Through a series of extrusion and size exclusion steps, CDNs are produced from in vitro cell culture and are found to highly resemble the endogenous exosomes. Thus, we envision that this strategy holds great potential as a viable alternative to exosomes in the development of ideal DDS.

Key words Drug delivery systems, Exosomes, Exosome-mimetics, Intrinsic targeting, Biomimicry

1 Introduction

Drug Delivery Systems (DDSs) include a wide range of technologies and/or devices that aim to deliver therapeutic cargos at the target site [1]. By increasing the accumulation of bioactive molecules at the diseased areas and/or reducing nonspecific exposure of the cargos to off-target sites, DDSs are expected to improve both efficacy and safety profiles of drugs.

Among the different formulations that have been proposed as suitable DDSs, the lipid-based nanovesicles called liposomes have always been the forerunner in the field of drug delivery. The main reason behind this is probably due to their lipid bilayer and aqueous core (which enable the encapsulation of both hydrophilic and lipophilic drugs), as well as their surface, which can be easily functionalized with targeting agents and/or imaging probes. To date, there are more than 50 FDA approved, commercially and clinically available liposomal formulations (e.g., Doxil®, Myocet®, and

Kumaran Narayanan (ed.), *Bio-Carrier Vectors: Methods and Protocols*, Methods in Molecular Biology, vol. 2211,
https://doi.org/10.1007/978-1-0716-0943-9_11,

Ambisome®) [2]. Of note, liposomes are one of the most extensively studied DDSs among the synthetic DDSs (e.g., polymeric micelles, metal nanoparticles, and carbon nanotubes), with well-established production protocols, drug loading procedures, and physical characterization methods.

This might seem surprising, since the mechanism by which liposomes accumulate at the target sites (such as tumor or inflammatory sites) is in common with other nanoscaled DDSs, and based on the so-called Enhanced Permeability and Retention (EPR) effect. This phenomenon, characterized by the "leaky" blood vessels with increased vascular permeability (with fenestrations between 0.2 and 1.2 μm) [3] at the diseased area, in contrast to the tight endothelial junctions of normal vessels (only up to about 10 nm) [4], is exploited by the nanometric dimensions of the nanocarriers (irrespective of the material used) and their ability to remain in the bloodstream for long time, rather than depending on the specific nature of the liposomes. Hence, liposomes do not possess any specific intrinsic targeting capability; on the contrary, they need to be further decorated at their surface (usually with poly (ethylene) glycol (PEG)) to avoid recognition and premature clearance by the mononuclear phagocyte system (MPS). Furthermore, not all target sites are susceptible to the EPR effect, as is the case of low-EPR tumors (e.g., pancreatic and bladder cancers) [5] and some target sites in the poor perfusion regions of the body [6]. Additionally, the inter- and intra-patient heterogeneities observed in the clinical settings also limit the effectiveness of this "passive targeting" (a.k.a. EPR-mediated targeting). Although much effort is devoted to employ ligands (e.g., peptides or antibodies) at the surface of nanoparticles to confer certain specificity through "active targeting," there has been only limited clinical success [7, 8].

The complex issue surrounding the practicality of nanoparticles as DDSs extends to the limited cellular uptake of liposomes by target cells or the potential elicitation of immune responses (e.g., complement activation-related pseudoallergies (CARPA) attributed to the synthetic lipid components or accelerated blood clearance (ABC) related to the generation of specific antibodies), which decrease the efficacy of liposomes upon repeated use in the clinical setting [9].

In recent years, cell-derived DDSs or extracellular nanovesicles have attracted growing interest and have been increasingly explored as novel platforms for drug delivery due to their intrinsic targeting ability [10, 11] . Among all cell-derived DDSs, exosomes (i.e., nanovesicles secreted endogenously by cells) are the most promising candidates due to their nanodimension (with size <300 nm, which enables them to exploit the EPR effect) [6, 12] and their recently discovered role in cell-to-cell communication (which is associated with a specific cellular uptake) [13, 14].

Exosomes are naturally occurring extracellular nanovesicles with a size range of 50–150 nm, and they are released by nearly all cell types [15, 16]. Their biogenesis involves the endosomal system and the formation of MultiVesicular Bodies (MVBs), where exosomes mature and are enriched with key surface protein markers (such as the endosomal sorting complexes required for transport (ESCRT)-associated proteins (Alix and TSG101), heat shock proteins (Hsp70 and Hsp90), and tetraspanins (CD9, CD63, and CD81)) prior to their release into the extracellular space [17–19].

As mediators of cell-to-cell communication in many physiological or pathophysiological conditions, exosomes are capable of shuttling many biomolecular cargos (such as proteins, carbohydrates, lipids, and nucleic acids) intercellularly. Furthermore, due to their natural origin, exosomes are proven to be much safer and more biocompatible DDSs as compared to their synthetic counterparts [20]. Owing to the surface molecules (mainly proteins and lipids) preserved from the parent cells, exosomes exhibit cell-specific targeting phenomena on certain recipient cells through adhesion and subsequent uptake (by fusion or endocytosis/phagocytosis) [19, 21, 22], which highlights the possibility of employing exosomes for targeted delivery. Indeed, exosomes have been explored for loading of various therapeutics, including small molecules (e.g., paclitaxel [23], doxorubicin (Dox) [24], and curcumin [25]), protein-based drugs (e.g., catalase [26]), and nucleic acids (e.g., siRNA [27] and miRNA [28]).

Up to date, exosomes have been extensively investigated clinically. At the time of publication, there are 124 clinical studies listed in US-NIH clinical trial database (https://clinicaltrials.gov/) with the keyword search of "exosomes." Although most of the clinical studies are associated with biomarker identification and diagnosis/prognosis of various diseases, we have summarized the studies related to the use of exosomes as DDSs in Table 1.

Despite the promising results in vitro and in vivo, the clinical translation of exosomes has been impeded by the inefficient extraction and tedious isolation process [20, 29]. Most of the exosome extraction methods are complex, labor-intensive, and time-consuming, yet the production yields are poor (Table 2). One of the most common methods for the isolation of exosomes involves ultracentrifugation or/and purification with sucrose density gradient. This process is time-consuming and requires large amount of starting materials (e.g., cell culture medium and cells) to be able to obtain sufficient amount of exosomes for cellular assays or in vivo applications (e.g., 10^6 cells can only produce approximately 0.5 μg of exosomes (in terms of protein content) [30]). Other methods, such as gel-filtration, polymeric precipitation, or immunoaffinity capture, have been reported to overcome the low production

Table 1
List of the clinical trials involving the use of exosomes as DDSs (https:/clinicaltrials.gov/)

NCT number	Study title	Phase	Condition	Origin of exosomes	Therapeutic cargo(s)
01294072	Study investigating the ability of plant exosomes to deliver curcumin to normal and colon cancer tissue	Phase I	Colon cancer	Plant	Curcumin
03608631	iExosomes in treating participants with metastatic pancreas cancer with KrasG12D mutation	Phase I	Pancreatic cancer	Mesenchymal stem cells	Small interference RNA (siRNA) against KrasG12D
01159288	Trial of a vaccination with tumor antigen-loaded dendritic cell-derived exosomes	Phase I	Non-small cell lung Cancer	Dendric cells	Tumor antigen
03384433	Allogenic mesenchymal stem cell-derived exosome in patients with acute ischemic stroke	Phase I Phase II	Acute ischemic stroke	Mesenchymal stem cells	miR-124
02565264	Effect of plasma-derived Exosomes on cutaneous wound healing	Early phase I	Cutaneous wound healing	Plasma	–
02138331	Effect of microvesicles and exosomes therapy on β-cell mass in type I diabetes mellitus (T1DM)	Phase II Phase III	Type I diabetes mellitus (T1DM)	Mesenchymal stem cells	–

yield and/or provide better isolation and purification [31–34]. Yet, by far, none of these methods is able to significantly increase the yields or reduce the isolation time of exosomes.

Cell-Derived Nanovesicles (CDNs) offer a promising alternative to exosomes. CDNs are produced by subjecting cells to serial extrusions and purification using size exclusion columns. Hence, their production is based on the principle of biomimicry of exosomes. As exosome-mimetics, CDNs were postulated to inherit the intrinsic properties of exosomes (namely size and targeting ability, biocompatibility, and non-immunogenicity), while being able to circumvent the poor production yields and lengthy procedures. Through the preservation of the membrane structures and components from the original parent cells, CDNs were hypothesized to be able to endow improved cellular uptake and reduced risk of immunogenicity similarly to endogenously produced exosomes, in comparison to synthetic liposomes.

Table 2
Summary of different methods for exosomes isolation, in terms of respective relative yield, processing time, advantages, and disadvantages [31–33]

Different methods (Example of the commercial kits)	Yield	Time required (h)	Advantages	Disadvantages	Ref
Ultracentrifugation	Low	3–4	High capacity	Time-consuming, low purity	[32, 57]
Gel filtration, e.g., *iZON, qEVSingle; 101Bio, PureExo kit*	Low	1.5–2	High purity, size uniformity	Low capacity, high cost	[32]
Immunoaffinity capture, e.g., *Wako, MagCapure*	Low	4–5	High purity, target specificity	Low capacity, high cost	[32, 57]
Polymertic precipitation,e.g., *SBI, Exoquick; Invitrogen, 4478359*	High	12–16	Simple procedure, high yield, size uniformity	Low purity, high cost	[32, 57]

A variety of methods have been empirically investigated for the production of CDNs, including the use of microfluidic channels [35, 36], extruders [37], and customized centrifugal devices [38] (Fig. 1). These methods were able to increase the production yield of exosome-mimetics up to 15 folds (Table 3) within a short time frame, compared to the exosomes isolated by traditional ultracentrifugation (from the same starting number of cells).

As reported by Jo et al. [38], the production of CDNs can be scaled up through the use of a customized centrifugal device. Our group developed a polycarbonate membrane fitted centrifugal device that is compatible to the common centrifuge, and demonstrated the ability of producing larger quantity of CDNs by increasing the number of cells used in the production (Fig. 3).

In addition, by using these physical extrusion methods, the production of CDNs offers new opportunities to perform surface functionalization or loading of therapeutic cargos. Jang et al. [37] further improved the production and loading of exosome-mimetic nanovesicles (NV) using a serial extrusion method. Common chemotherapeutics, such as doxorubicin, carboplatin, and gemcitabine, have been successfully loaded into these NVs (Fig. 2). These loaded NVs displayed better selectivity and concentration-dependent cytoxicity at the diseased areas, and it was observed that doxorubicin-loaded NVs were more potent and cytotoxic on TNF-α-treated HUVEC cells (which imitated the endothelial cells at inflammation site) than free doxorubicin at the same dose of 1.5 μg/mL doxorubicin [37].

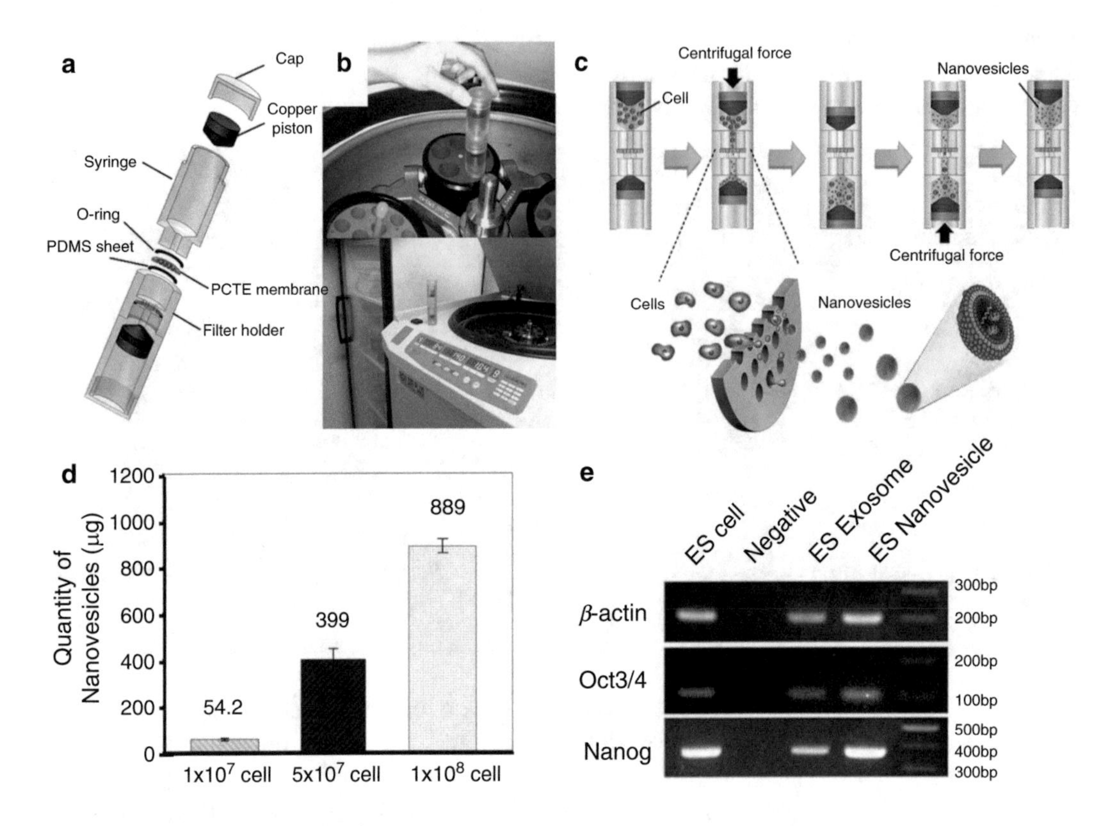

Fig. 1 Production of cell-derived nanovesicles from murine embryonic stem (ES) cells. (**a**) Design of the centrifugal device. (**b**) Photograph of the device in operation. (**c**) Schematic outline of the process for nanovesicles production. (**d**) amount of nanovesicles generated with different initial cell numbers. (**e**) Reverse transcription-PCR for ES cells, exosomes, and nanovesicles generated by 1×10^8 ES cells. (Reproduced from Jo et al. (2014) [38] with permission from The Royal Society of Chemistry)

However, despite the increased production yield, most of the CDNs/NVs production technologies are not commonly used in the laboratory due to the necessity of highly specialized and expensive equipment with customization (e.g., microfluidic channels and customized device). Therefore, Goh et al. recently developed a simple, rapid, and cost-effective method for the production of CDNs using readily available equipment and setups: a bench-top microcentrifuge with temperature control and spin cups fitted with membranes with various pore sizes [29]. Compared to other reported protocols, this method has simplified the workflow of CDNs production. Through the use of common equipment and consumables, this production method can be easily adapted and performed in most laboratories.

In this chapter, we present the protocol adopted from Goh et al. [29], where CDNs were produced from centrifugation-shearing methods (Fig. 3), by subjecting the cells through serial centrifugation steps to produce nanovesicles that resemble endogenous exosomes both physically and biochemically.

Table 3
Production yield (in terms of protein amount) of exosomes and exosome-mimetics using different methods. Cell lines indicated in bold are derived from cancer

Types of vesicles	Method of production	Type of cells	Protein amount (μg, normalized to 2×10^7 starting cell number)	Ref.
Exosomes	Ultracentrifugation	Mouse dendritic cell	~10	[30]
		Natural killer cells	~6	[58]
		U937	~40	[29]
		B16BL6	~1000	[59]
		MiaPaCa cells	~460	[32]
		Murine embryonic stem cells	~4	[38]
	Gel filtration	**MiaPaCa cells**	~600	[32]
	Immunaffinity capture	**MiaPaCa cells**	~160	[32]
	Polymeric precipitation	**B16BL6**	~2200	[59]
		MiaPaCa cells	~1633	[32]
Exosome-mimetics	Extrusion	**Natural killer cells**	~74	[58]
		U937	~406	[37]
	Customized centrifugal device	Murine embryonic stem cells	~184	[38]
	Microchannel	Murine embryonic stem cells	~270	[35]
			~500	[36]

U937 cells (human monocytes) were used as the model for CDNs production in this protocol. CDNs derived from U937 were demonstrated to inherit the intrinsic targeting ability towards inflammatory sites (e.g., tumor) [37] from the parent monocytes [39].

Comparative analyses between CDNs and exosomes from U937 were performed to assess the feasibility of the CDNs production method in terms of physical characteristics, production yield and duration, protein markers, and lipid contents. The functionality of the CDNs was also demonstrated through in vivo biodistribution studies in a mouse CT26 xenograft model.

The CDNs produced through this protocol had similar physical characteristics as exosomes in terms of hydrodynamic size, zeta potential, and morphology (Fig. 4a–c), suggesting that this method is capable of improving the production yield and reducing the processing time, while preserving the physical properties of exosomes. With the same amount of starting cells (i.e., 2×10^7 cells/mL), the yield of CDNs (in terms of protein

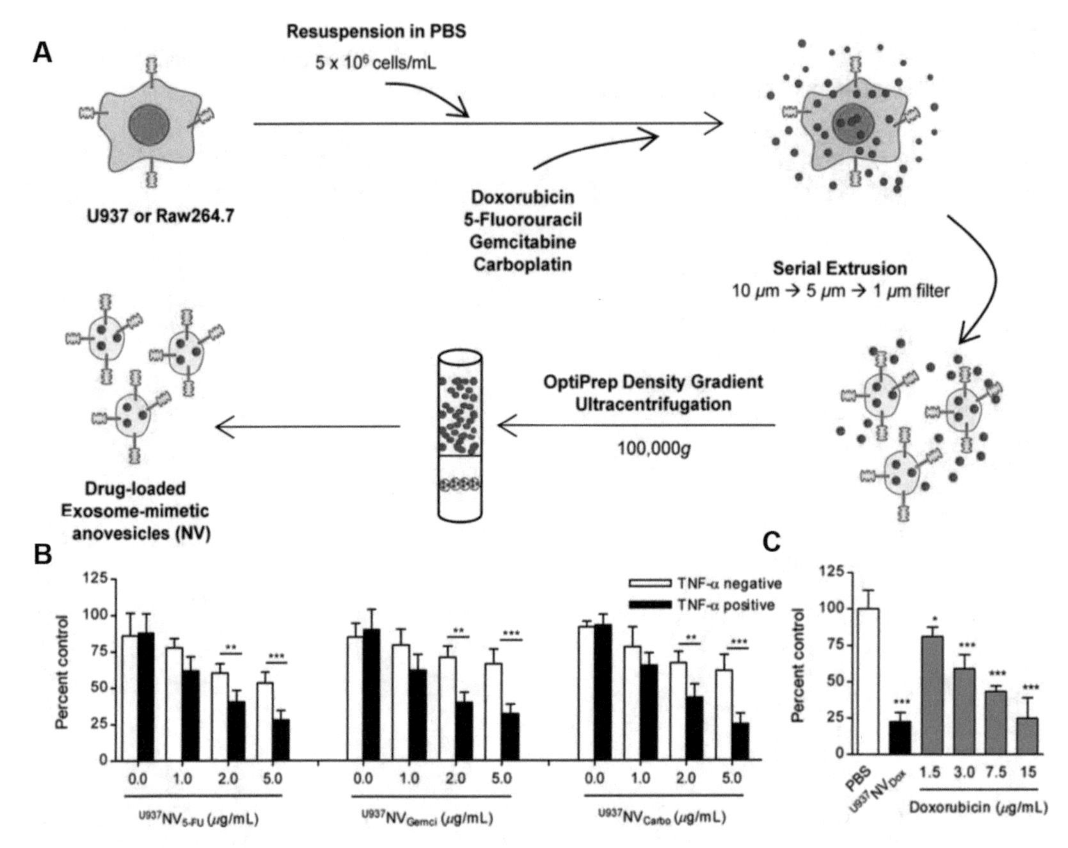

Fig. 2 Production of exosome-mimetic nanovesicles (NV) and chemotherapeutics-loaded NV and their in vitro targeted delivery. (**a**) Schematic illustration of the procedure for the generation of nanovesicles (NV) and chemotherapeutics-loaded NV. (**b**) Cytotoxic effects of various chemotherapeutics-loaded NV on TNF-α-treated and untreated HUVECs ($n = 6$/group). (**c**) Comparison of the cytotoxic effects of doxorubicin-loaded NV with varying doses of free doxorubicin on TNF-R-treated HUVECs ($n = 6$/group). 1.5 μg of doxorubicin is loaded into the $^{U937}NV_{Dox}$ (5 μg of total protein). Data are presented as the mean (SD $*P < 0.05$, $**P < 0.01$, $***P < 0.001$. *EXO* exosomes, *Dox* doxorubicin, *Gemci*gemcitabine, *Carbo* carboplatin. (Adapted from Jang et al. [37] with permission from the American Chemical Society)

concentration = 540 μg/mL) (Fig. 4d) was about 15-fold higher than that of the exosomes isolated using ultracentrifugation (40 μg/mL), and was also higher than other reported methods (about 406 μg/mL from a cell density of 2×10^7 cells/mL [37]).

Besides having a higher yield, this approach also presented a higher time-saving efficiency of up to 3 days, as compared to the traditional exosome isolation method using ultracentrifugation (Fig. 4e) In this method, CDNs were obtained in less than a day after harvesting the cells, whereas exosomes from ultracentrifugation can take up to 4 days with the time-lag between harvest of cells and secretion of exosomes.

The classic exosome protein markers, namely tetraspanins (CD9) and MVB markers (Alix and TSG101), were used to evaluate the resemblance of the CDNs with exosomes. Our results

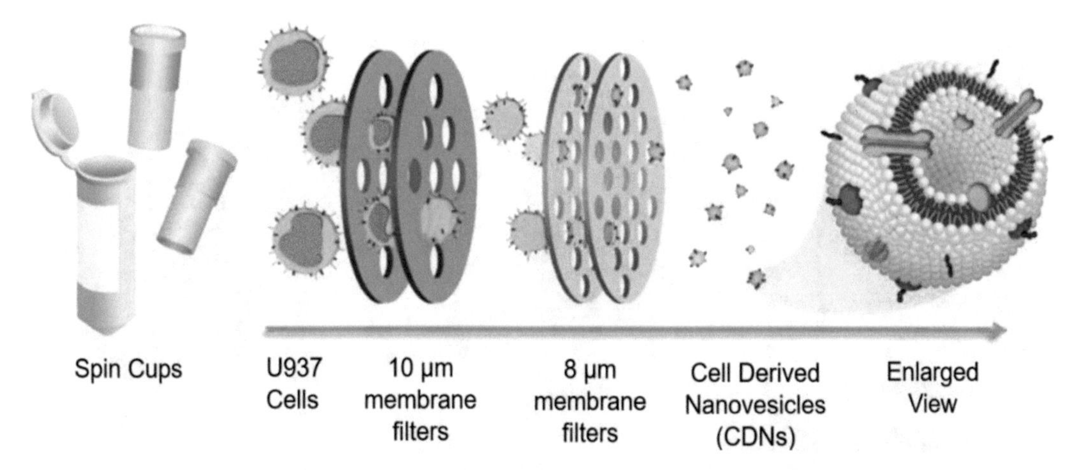

Fig. 3 Schematic illustration of CDNs production. The U937 cells were passed through the spin cups fitted with various size membrane filters and subsequently purified via size extrusion to obtained CDNs with desired size. (Reproduced from Goh et al. [29] with permission from SpringerNature)

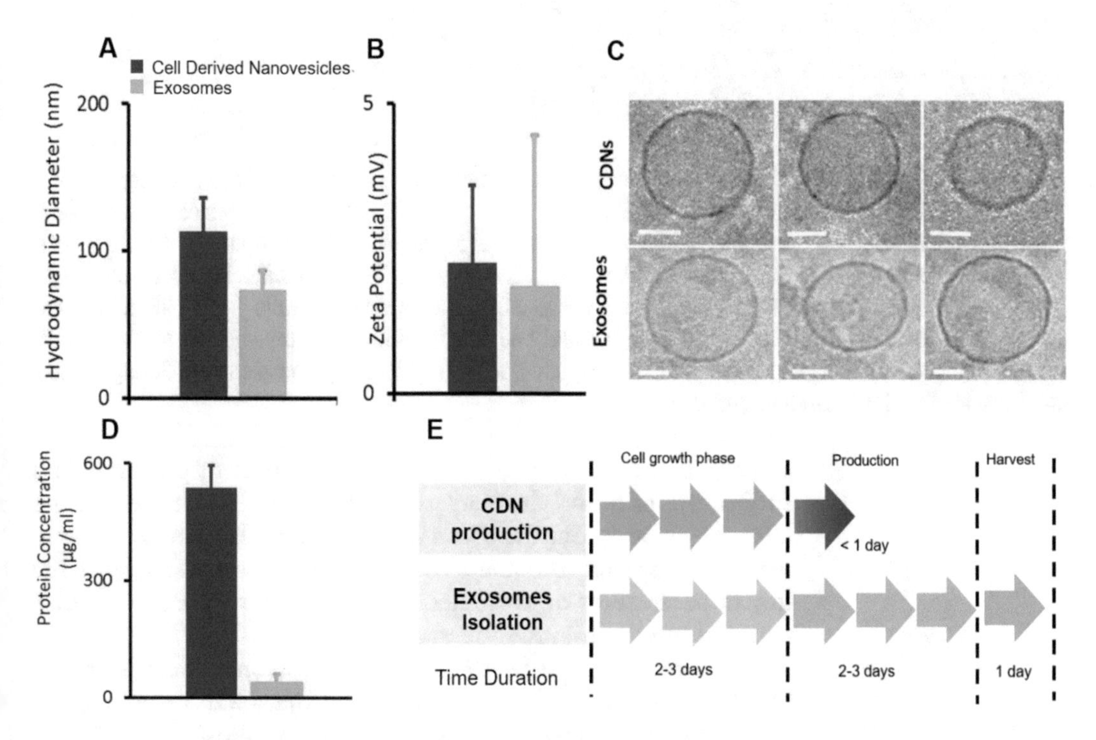

Fig. 4 Physical characterization of CDNs and comparison of production processes of exosome isolation and CDNs production. (**a**) Hydrodynamic diameter and (**b**) Zeta potential of CDNs and exosomes. (**c**) Cryo-TEM images of CDNs (top row) and exosomes (bottom row). Scale bars indicate 20 nm. (**d**) Production yield of CDNs and exosomes in terms of protein concentration. (**e**) production time of CDNs and exosomes per production run. Data represent means ± SEM ($n = 3$). (Adapted from Goh et al. [29] with permission from SpringerNature)

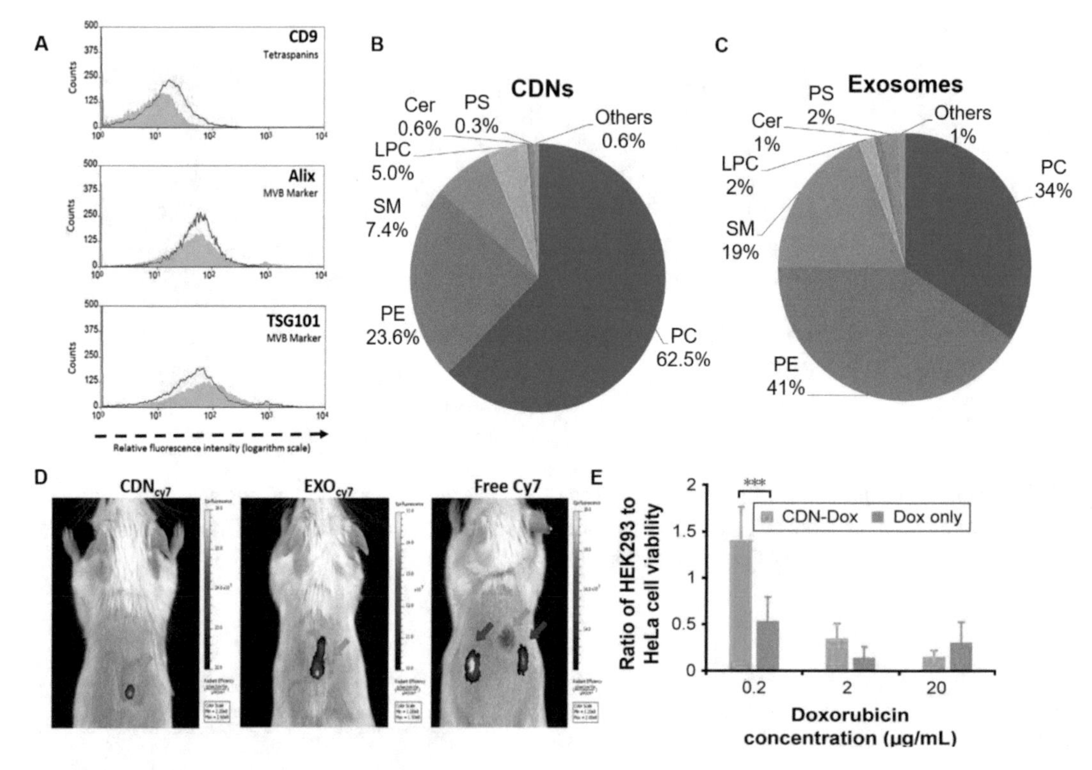

Fig. 5 (**a**) Histogram plots of three key protein markers (namely tetraspanins: CD9, MVB markers: Alix and TSG101) in CDNs and isolated exosomes obtained using flow cytometry. CDNs and exosomes are depicted in green and red, respectively. Lipidomic profiles of (**b**) CDNs and (**c**) exosomes derived from 2×10^7 U937 cells. (**d**) Accumulation of Cyanine7-labeled CNDs, exosmes, and free dye in the mice xenograft model. Tumor sites are denoted by green arrows and the kidneys are denoted by red arrows (**e**) The ratio of HEK293 to HeLa cell viability in co-cultures, incubated with CDNs-Dox and free Dox for 48 h. Data represented means $\pm$ SEM ($n = 3$) ***$p < 0.001$. (Adapted from Goh et al. [29] and Goh et al. [40] with permission from SpringerNature and Dove Medical Press Limited, respectively)

showed that all three protein markers were present on the CDNs surfaces, confirming the similarity between CDNs and exosomes (Fig. 5). In addition, the preservation of these key surface protein markers suggested the ability of CDNs to retain and inherit the functional features of exosomes, such as targeting ability, biocompatibility, and non-immunogenicity.

The lipidomic profile of CDNs and exosomes were also analyzed and compared. The data indicated that CDNs were able to retain the major lipid components (namely phosphatidylcholine (PC), phosphatidylehanolamine (PE), sphingomyelin (SM), and lysophosphatidylcholine (LPC)) found on exosomes (Fig. 5b, c). Since the lipid components of the vesicles are one of the major players defining the physical characteristics such as fluidity, permeability, and stability of the lipid bilayer that forms the vesicles, these similar lipid contents further corrobarated the analogy of CDNs to exosomes.

Similar accumulation profiles between CDNs and exosomes were observed in an in vivo biodistribution study using mouse xenograft model of colon carcinoma (Fig. 5d), indicating that CDNs were able to mimic the endogenous exosomes [29]. These data confirm the ability of CDNs and exosomes to accumulate at the diseased area (in this case the tumor tissue). On the other hand, we also confirmed the intrinsic targeting abilities of monocyte-derived CDNs towards cancer cells rather than healthy cells. It was observed that doxorubicin-loaded CDNs reduced the viability of HeLa cells (a cancer cell line) more than the co-cultured HEK293 cells (a non-cancer cell line). This is in contrast to the effect of free doxorubicin, where the cell viability of both cell lines was reduced to a similar extent. Indeed, the results further exemplify the cancer targeting potential of CDNs, in both monocultures and co-cultures, an effect we attributed to the retention of original monocytes' membrane surface from which the CDNs were derived [40].

All in all, though exosomes and exosome-mimetics hold tremendous potential in the development of novel targeting DDSs, there are challenges associated with drug loading, in vivo clearance, and possible immunogenic effects. First, relatively poor loading was observed when CDNs were loaded with doxorubicin (encapsulation efficiency (EE%) ~17%) [40], compared to the synthetic liposomes (with EE% for doxorubicin >90%) [42]. This has prompted researchers and scientists to extend their exploration and develop various types of bioinspired DDSs (Table 4). *Examples of bioinspired DDSs.*). Among them, hybrid systems obtained from the fusion of synthetic and cell-derived components have gained much interest in recent years, as a potential solution to circumvent the challenges faced by cell-derived DDSs. By combining the more established synthetic DDS with the nascent cell-derived DDS, these hybrid DDSs aim to exploit the benefits of both systems and to compensate on each other's short-comings. One of these hybrid DDSs includes EXOPLEXs, a chimeric system formed from fusion of CDNs and synthetic lipids (Fig. 6), which showed an EE% of doxorubicin >65% [40, 41]. EXOPLEXs were able to retain the characteristic protein markers of CDNs and doxorubicin-loaded EXOPLEXs displayed an improved cancer killing effect compared to doxorubicin-loaded liposomes.

Second, one of the lessons learnt from the in vivo study of unmodified exosomes was their rapid clearance: these unmodified exosomes were subjected to rapid clearance from the blood circulation (half-life about 2 min) upon intravenous administration. Studies found that these exosomes were rapidly sequested in spleen, liver, and lungs [43, 44] when administrated intravenously, whereas they were retained in the tumor when administrated intratumorly [45]. This suggests that, in order to exploit the intrinsic targeting abilitiy of exosomes and exosome-mimetics upon i.v. injection,

Table 4
Examples of bioinspired DDSs

Basis of DDS	Technology	Production strategies	Therapeutic cargo (s)	Ref
Cell-derived	Exosomes	Isolation	Paclitaxel	[23]
			Doxorubicin	[24]
			Curcumin	[25]
			Catalase	[26]
			Porphyrins	[60]
			siRNA	[27]
			miRNA	[28]
	Exosomes memetics	Cell shearing	Doxorubicin	[29, 37]
			5-Fluorouracil	[37]
			Gemcitabine	[37]
			Carboplatin	[37]
		Liposomes formulated with lipid composition of natural exosomes	siRNA	[61]
	Nanoghosts	Shearing of cell ghosts[a]	Plasmid DNA	[62]
			Doxorubicin	[63]
			–	[64]
	Cell-derived liposomes		EDTA	[65]
Hybrid system (cell-derived + synthetic system)	Engineered exosomes	Complex of exosomes with cationic lipid and pH-sensitive fusogenic peptide	Saporin	[66]
	Hybrid exosomes	Complex of exosomes with cationic lipid	Plasmid DNA	[67]
		Fusion of exosomes with various lipids	–	[68]
			mTHPC	[69]
	PEGylated exosomes	Post-insertion of PEG lipids to exosomes	–	[70]
	EXOPLEXs	Fusion of CDNs with neutal lipids	Doxorubicin	[41]
	Nano-cell vesicle technology system (nCVTs)	Fusion of cell ghosts[a] with neutral lipids	–	[71]
	Exosome-coated nanoparticles	Fusion/coating of exosome with the surface of nanoparticles	Suberohydroxamic acid	[72]
	Cell membrane-coated nanoparticles	Coating of cancer cell membrane with nanoparticles through coextrusion	–	[73, 74]

[a]Cell ghosts are cells that devoid of cellular contents

there is a need to engineer their surface (e.g., through PEGylation) for improvement of their circulation time and enhancement of their accumulation at the in vivo targets.

Third, another potential drawback is with regard to the immunogenicity. Though it has been claimed that endogenous exosomes are likely to exhibit less immunogenic potential compared to other

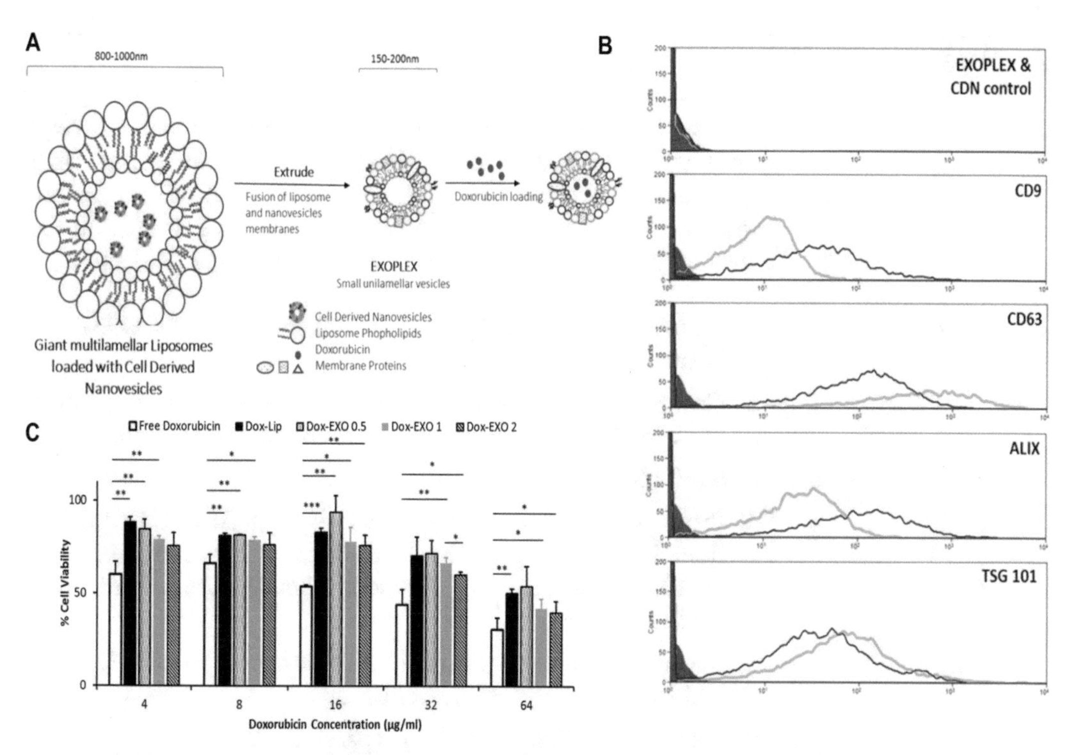

Fig. 6 Production of EXOPLEXs and their in vitro cytotoxicity. (**a**) Schematic illustration of EXOPLEXs production from fusion of liposomes and CDNs. (**b**) Flow cytometry analysis of characteristic CDNs protein markers on EXOPLEXs. Red regions indicate beads only, and the green and blue lines indicate CDNs and EXOPLEXs, respectively. (**c**) Cell viability assay on HeLa cells over 48 h. Data represented means SEM ($n = 3$) (SD $^{*}P < 0.05$, $^{**}P < 0.01$, $^{***}P < 0.001$). (Adapted from Goh et al. [41] with permission from the American Chemical Society)

synthetic counterparts owing to their autologous sources, little is known about their long-term immune stimulation and immunogenicity potential. Furthermore, the shearing process of CDNs production, while retaining crucial protein markers of endogenously produced exosomes, may potentially alter the membrane conformation and composition [38]. Nonetheless, further studies are necessary to identify the specific components responsible for the high organotropism of exosomes. Moreover, future development of exosome-mimetics aimed to overcome the drawbacks of the current DDSs will pave the way for the development of an ultimate safe and targeted delivery system.

In the following sections, we cover the step-by-step procedures for producing CDNs as reported by Goh et al. [29]. The rationale for each step is discussed in the "Notes" sections.

2 Materials

All reagents were prepared using Milli-Q water (obtained through purifying the deionized water to attain a resistance of 18 MΩ cm at 25 °C) and filtered through 0.22 μm (sterilization) prior to use. All biological wastes were discarded into the biohazard bag and disposed in accordance with the proper waste disposal guideline.

2.1 Cell Culture

1. Human pro-monocytic myeloid leukemia cell line (U937) (*see* **Note 1**).
2. Complete culture medium: RPMI-1640 medium supplemented with 10% heat-inactivated Fetal Bovine Serum (FBS).
3. Sterile 50 and 15 mL tubes.
4. Protease inhibitor cocktail (EDTA-free,): AEBSF hydrochloride (100 mM), aprotinin (bovine lung) (0.08 mM), bestatin: (5 mM), E-64 (1.5 mM), leupeptin hemisulfate (2 mM), pepstatin A (1 mM).
5. Phosphate-Buffered Saline (PBS): adjusted to pH 7.4 with hydrochloric acid (HCl), sterilized.
6. 0.4% Trypan Blue solution: prepared in PBS.
7. Light microscope.
8. Hemocytometer (cell counting chamber).

2.2 Centrifugation

1. Benchtop refrigerated microcentrifuge machine with the capacity of reaching 14,000 × *g*.
2. Spin cups supplied with 10 μm filters attached.
3. 8 μm Hydrophilic polycarbonate membranes.
4. Hole punctures (hole diameter: 9 mm).

2.3 Size Exclusion Chromatography (SEC)

1. Sephadex G-50 self-packed size exclusion column.
2. PBS: Ice-cold, pH 7.4.
3. Sterile 1.5 mL microcentrifuge tubes.

2.4 Physical Characterization

2.4.1 Hydrodynamic Size and Zeta Potential Measurement

1. CDNs samples.
2. Zetasizer Nano (Malvern Instruments).
3. Folded capillary cell/polycarbonate cell with gold-plated electrodes.
4. Disposable polystyrene cuvette.
5. De-ionized (D.I.) water.
6. Kimwipes.

2.4.2 Protein Quantification

1. CDNs samples.
2. Standard BCA assay kit.
3. PBS: pH 7.4.
4. 25–2000 μg/mL Bovine Serum Albumin (BSA) standards (dissolved in PBS).
5. Microplate reader with a 562 nm filter.
6. Clear Flat-bottom 96-well plate.
7. Micropipettes.

3 Methods

3.1 Cell Culture

1. Culture U937 cells in complete culture medium and maintain at 37 °C with 5% CO_2, until the cells reach 70–80% confluency (*see* **Note 2**).
2. Harvest the cells by centrifuging at 4 °C (500 × *g* for 10 min) (*see* **Notes 3** and **4**).
3. Remove the supernatant (*see* **Note 5**).
4. Resuspend the cells in fresh complete culture medium and collect an aliquot for cell number determination using the standard cell counting chamber (*see* **Note 6**).
5. Pipette the amount containing 2×10^7 cells into a fresh sterile 15 mL centrifuge tube (*see* **Note 7**).
6. Collect the cells by centrifugation at 4 °C (500 × *g* for 10 min) and discard the supernatant (*see* **Notes 3** and **4**).
7. Wash U937 cells with sterile PBS twice by resuspending the cell pellet with PBS and repeating the centrifugation step to remove any residual cellular debris, dead cells, and cell culture medium (*see* **Note 8**).
8. Resuspend the cells in 1 mL of sterile PBS with protease inhibitor cocktail (1:200 dilution) (*see* **Note 9**).
9. Use cells immediately, otherwise leave them on ice and use within 1 h (*see* **Note 4**).

3.2 Cell Shearing Via Centrifugal Force

The centrifugal force was used in the cell disruption process to shear the U937 cells to produce the CDNs.

1. Pre-cool the microcentrifuge to 4 °C (*see* **Notes 3** and **4**).
2. Obtain 10 pieces of 8 μm membrane filters (9 mm diameter) using the hole puncher (*see* **Note 10**).
3. Place two pieces of 8 μm membrane into each spin cup (*see* **Note 10**).

4. Divide the 1 mL cell suspension with 2×10^7 cells into 5 aliquots (200 μL/each).
5. Place each aliquot in a spin cup fitted with a 10 μm membrane filter and centrifuge at 14,000 × *g*, 4 °C for 10 min. (*see* **Notes 3** and **4**).
6. Re-introduce the flow-through to the same spin cup and repeat the centrifugation step (**step 5**).
7. Pipette the flow-through to another spin cup fitted with two 8 μm membrane filters and centrifuged 14,000 × *g* for 10 min.
8. Re-introduce the flow-through to the same spin cup with 8 μm membrane filters and centrifuge 14,000 × *g* for 10 min.
9. Collect the flow-through of each spin cup (*see* **Note 11**).

3.3 Purification Process

1. Pool the flow-through from five different spin cups to obtain 1 mL of the raw CDNs sample.
2. Equilibrate the Sephadex G-50 Size Exclusion Column (SEC) with ice-cold PBS.
3. Add the raw CDNs onto SEC slowly to prevent disturbance of the Sephadex G-50 beads.
4. Collect each 0.5 mL (one fraction) in a new 1.5 mL microcentrifuge tube. Fraction 3 and 4 are CDNs (*see* **Note 12**).
5. Pool the fractions (fraction 3 and 4) and store on ice temporarily for the characterization and further processings or at 4 °C for up to 3 days (*see* **Note 13**).

3.4 Characterization

3.4.1 Hydrodynamic Size Measurement

The following protocol described the steps of using a Malvern (Nano series) instruments. The protocol may need slight modifications for the application in other systems. The Zetasizer Nano instrument should be switched on for at least 30 min before the measurement to allow the laser to warm up.

1. Pipette the 50 μL of the sample into a 1.5 mL microcentrifuge tube.
2. Dilute the sample with 450 μL of D.I. water (10× dilution) (*see* **Notes 14** and **15**).
3. Transfer the diluted sample into the cuvette.
4. Insert the cuvette into the instrument (*see* **Note 16**).

3.4.2 Zeta Potential Measurement

1. Pipette 75 μL of the sample into a microcentrifuge tube.
2. Dilute the sample with 675 μL of D.I. water (10× dilution) (*see* **Notes 14** and **15**).
3. Slowly inject the sample into the cuvette to prevent the introduction of air bubbles.
4. Insert the cuvette into the instrument (*see* **Note 16**).

3.4.3 Protein Quantification Assay

The following protocol described steps of using a Pierce™ BCA Protein Assay Kit. The protocol may need slight modifications for the application in other assays (*see* **Note 17**).

1. Pipette 25 μL of the CDNs samples into a 1.5 mL microcentrifuge tube.
2. Dilute the samples with 25 μL of PBS (2× dilution), and stored on ice prior to the assay.
3. Prepare BSA standards, which can be stored at 4 °C for up to 2 weeks.
4. Prepare BCA reagent. Mix it well and protect from light.
5. Load 25 μL of the standards and samples into the wells of a transparent 96-well plate.
6. Add 200 μL of BCA reagent into each well.
7. Incubate at 37 °C for 30 min prior to the analysis at 562 nm using microplate readers.

4 Notes

1. U937 is a suspension cell line. There is no need for enzymatic treatment (trypsinization) to detach cells from the surface of cell culture flask during the subculture or passage. Instead, cells are continuously maintained by diluting the cells with fresh culture medium.
2. Cells (either suspension or adherent cells) should be grown to reach 70–80% confluency. It is important to ensure the cells are not over confluent, as it may lead to changes in expression of cell surface protein/markers [46] and gene expression [47, 48]. Furthermore, when the cells deplete all the nutrients supplemented in the culture medium, they may cease proliferation and enter quiescence (reversible cell cycle arrest), then eventually become senescent (irreversibly lose cellular functions).
3. All centrifugations steps should be performed at 4 °C. All the tubes should be weighed before the centrifugation to ensure the weight balance during the centrifugation.

   The cold condition is important to slow down the enzymatic reaction of intracellular proteases, thus preventing any degradation and proteolysis.
4. The supernatant should be removed from the tube carefully in one smooth motion to prevent the disturbance of the cell pellet. The supernatant consists of spent medium, cellular debris, and dead cells. The presence of dead cells and debris may negatively affect the health of other viable cells, by causing

clumping or inducing cell death. If adherent cells are used for CDNs production, the cells should be rinsed with PBS before trypsinization and collected into a fresh sterile tube, followed by the same production protocol.

5. Trypan blue staining allows the discrimination between the viable cell and dead cells: as the dead cells are unable to exclude the dye, they become stained in blue. The trypan blue solution should be filtered to remove any particles or precipitates in the solution, which may interfere with the counting. A pipette is used to fill the hemocytometer chamber slowly and continuously to prevent any bubble formation in the chamber. The total number of cells and blue staining cells are counted under the light microscope. Cell viability can be calculated by the following equation:

$$\%\text{viable cells} = \left(1 - \frac{\text{Number of blue cells (dead)}}{\text{Total number of cells}}\right) \times 100$$

   The healthy log-phase cell culture should have at least 95% cell viability.

   For a more accurate measurement, a total number of cells counted within each square should be within 50–100 cells. If the cell count is more than 200 cells/square, the cell aliquot should be further diluted before repeating the cell counting process.

6. The hydrodynamic size of the CDNs was found to change when different cell densities were used for CDNs production [29]. While the exact optimal cell density can be determined empirically, one needs to take the diameter of the membrane used (in the spin cup) into consideration, as it was hypothesized that there is a relationship between the available surface area (for cells to pass through) and the number of cells. Thus, the number of cells used for CDNs production may affect the amount of cell-to-cell attrition and subsequently the size of CDNs. The cell density will also be different when using different cell lines for production of CDNs. The exact optimal cell density should be determined empirically. The general guideline is as follows: when using cells with small size (~10 μm) (e.g., RAW 264.7 or U937), higher starting cell number (~2×10^7 cells/mL) should be considered, whereas less starting cell numbers may be used when the cell size is larger (>20 μm) (e.g., 3T3-L1 or HEK293).

7. FBS is commonly used as a growth supplement in most of the cell culture media. It provides essential nutrients and growth factors that enable the proliferation of cells in vitro [49]. FBS is a serum product from cow, thus containing a significant amount of cow-derived extracellular vesicles (including exosomes). Their presence can potentially interfere with the

exosomes isolation and lead to a false high protein amount in the downstream analysis [50].

Similarly, in the case of CDNs production, several washing steps are needed during cell harvesting to remove any residual FBS on the cells prior to the centrifugation. This is to avoid the interference of the cow-derived exosomes when quantifying CDNs.

8. Cells contain a large number of proteases. During the shearing process (centrifugation), as the homeostasis and the controlled environment become unbalanced, the endogenous proteases could potentially degrade the surface membrane proteins on CDNs [51]. The proteases used in this protocol include a cocktail, containing protease inhibitors for serine proteases, aminopeptidase B, leucine aminopeptidase, cysteine proteases, and aspartic proteases. The detailed concentration can be found at https://www.abcam.com/protease-inhibitor-cocktail-edta-free-ab201111.html
9. Use the regular hole puncher to cut the 8 μm membrane filter into 9 mm size rings. Two pieces of the 8 μm membrane are carefully placed into the spin cups using the tweezers as shown in Fig. 7. The two membrane should completely overlap with each other.
10. The collected flow-through contains cell-derived vesicles and larger cell debris. Hence, further purifications using SEC is required to obtain CDNs with homogeneous size. It is

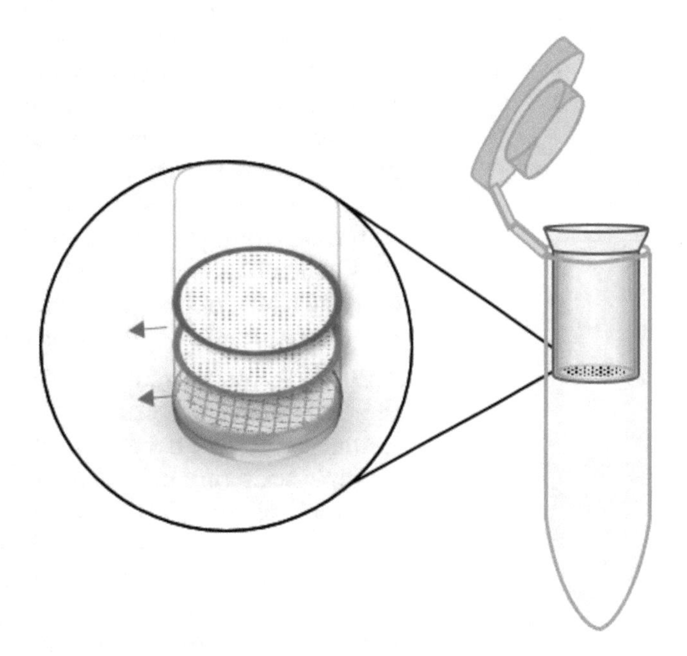

Fig. 7 Schematic illustration of how the 8 μm membrane is placed into the spin cup

important to separate the CDNs from the cell debris and other interfering components to ensure the reliability and accuracy of the downstream application [52]. Sephadex G-50 column will separate the particles with different sizes and molecular weights (size-exclusion) [53]; thus, this column is able to separate the CDNs with the desired hydrodynamic size from other bigger or smaller cellular debris or vesicles.

11. It is advisable to collect all fractions for protein quantifications and physical characterizations for the first time of CDNs production. CDNs should possess relatively high protein amount and appropriate hydrodynamic diameters.
12. The stability of CDNs were examined at 4 and 25 °C. The size of the exosomes was maintained around 150 nm at 4 °C for up to 7 days, whereas the size of CDNs increased drastically after 1 day stored at 25 °C, indicating the instability due to possible structural changes or degradation [29]. Our group has successfully performed the lyophilization of CDNs, as it is a commonly used technique for preservation and long-term stability of biological samples, such as proteins and exosomes. It was shown that lyophilized CDNs preserved physical characteristics and functionality (unpublished results), which further corroborated the versatility of this CDNs production protocol.
13. The selection of dispersant is important for zeta potential measurement, as the zeta potential measurement is affected by the ionic strength and pH of the surrounding medium [54, 55]. The rule of thumb for choosing the diluent for the sample is to always perform a size measurement prior to the zeta potential measurement. In this way, it is ensured that the particles of interest are stable in the chosen buffer and no aggregation or degradtion occur. For any physiological application, it is advisable to choose the buffer to maintain the pH at the physiological range (pH 7.4) for a more accurate depiction of the behavior of particles in physiological conditions. 0.1× PBS (10 mM PBS) is usually recommended for the routine analysis. Thus, in this protocol, we performed a 10× dilution to our samples (in 1× PBS) for the measurement.
14. The sample concentration may affect the size measurement. At the low sample concentration, the light scattering may be insufficient for the detection, while, at the high concentration, multiple scattering may happen, leading to inaccuracy for the size measurement [56]. It is advisable to keep the mean count rate of the samples between 200 and 500 kilo counts per second (kcps) to ensure an accurate measurement. For more details on the setting of the measurement protocol, please refer to the Malvern user manual. (https://www.malvernpanalytical.com/en/learn/knowledge-center/user-manuals/MAN0485EN

15. Follow the instruction of the manufacturer on how to insert the cell/cuvette into the machine. For detailed instructions for different types of cuvettes, please refer to the Malvern user manual. (https://www.malvernpanalytical.com/en/learn/knowledge-center/user-manuals/MAN0485EN)
16. We used the BCA assay to indirectly quantify the amount of surface proteins present on the CDNs [38], which serves as an indicator for the quantity of CDNs. The detailed instruction of how to perform the assay can be found at the supplier's website. (https://www.thermofisher.com/order/catalog/product/23225)

Acknowledgments

The authors acknowledge the National University of Singapore (NUS) for financial support (grant C-141-000-097-001), A-STAR SERC for financial support (grant number: 152 80 00046 (R-148-000-222-305)), IAF-PP (grant number A20G1a0046, R-148-000-307-305) and the Ministry of Education Singapore (MOE) for financial support (grant R148-000-267-114, R-148-000-284-114 and R-148-000-296-114).

References

1. Tiwari G, Tiwari R, Sriwastawa B, Bhati L, Pandey S, Pandey P, Bannerjee SK (2012) Drug delivery systems: an updated review. Int J Pharm Investig 2:2–11
2. Bulbake U, Doppalapudi S, Kommineni N, Khan W (2017) Liposomal formulations in clinical use: an updated review. Pharmaceutics 9. https://doi.org/10.3390/pharmaceutics9020012
3. Desai N (2016) Nanoparticle albumin-bound paclitaxel (Abraxane®). In: Albumin med. Springer Singapore, Singapore, pp 101–119
4. Sukriti S, Tauseef M, Yazbeck P, Mehta D (2014) Mechanisms regulating endothelial permeability. Pulm Circ 4:535–551
5. Zhang B, Hu Y, Pang Z (2017) Modulating the tumor microenvironment to enhance tumor nanomedicine delivery. Front Pharmacol 8:952
6. Kobayashi H, Watanabe R, Choyke PL (2013) Improving conventional enhanced permeability and retention (EPR) effects; what is the appropriate target? Theranostics 4:81–89
7. Rosenblum D, Joshi N, Tao W, Karp JM, Peer D (2018) Progress and challenges towards targeted delivery of cancer therapeutics. Nat Commun 9:1410
8. Belfiorea L, Saundersb DN, Ransona M, Thurechtc KJ, Stormd G, Vinea KL (2018) Towards clinical translation of ligand-functionalized liposomes in targeted cancer therapy: challenges and opportunities. J Control Release 277:1–13
9. Sercombe L, Veerati T, Moheimani F, Wu SY, Sood AK, Hua S (2015) Advances and challenges of liposome assisted drug delivery. Front Pharmacol 6:286
10. Li Y-J, Wu J-Y, Hu X-B, Wang J-M, Xiang D-X (2019) Autologous cancer cell-derived extracellular vesicles as drug-delivery systems: a systematic review of preclinical and clinical findings and translational implications. Nanomedicine 14:493–509
11. Chong SY, Lee CK, Huang C et al (2019) Extracellular vesicles in cardiovascular diseases: alternative biomarker sources, therapeutic agents, and drug delivery carriers. Int J Mol Sci 20:3272
12. Bunggulawa EJ, Wang W, Yin T, Wang N, Durkan C, Wang Y, Wang G (2018) Recent advancements in the use of exosomes as drug delivery systems. J Nanobiotechnol 16:81

13. Abels ER, Breakefield XO (2016) Introduction to extracellular vesicles: biogenesis, RNA cargo selection, content, release, and uptake. Cell Mol Neurobiol 36:301–312
14. Tricarico C, Clancy J, D'Souza-Schorey C (2017) Biology and biogenesis of shed microvesicles. Small GTPases 8:220–232
15. Pegtel DM, Peferoen L, Amor S (2014) Extracellular vesicles as modulators of cell-to-cell communication in the healthy and diseased brain. Philos Trans R Soc Lond B Biol Sci. https://doi.org/10.1098/rstb.2013.0516
16. Greening DW, Xu R, Ji H, Tauro BJ, Simpson RJ (2015) A protocol for exosome isolation and characterization: evaluation of ultracentrifugation, density-gradient separation, and immunoaffinity capture methods. Humana Press, New York, pp 179–209
17. Kumar D, Gupta D, Shankar S, Srivastava RK (2015) Biomolecular characterization of exosomes released from cancer stem cells: possible implications for biomarker and treatment of cancer. Oncotarget 6:3280–3291
18. Raposo G, Stoorvogel W (2013) Extracellular vesicles: exosomes, microvesicles, and friends. J Cell Biol 200:373–383
19. Van Niel G, D'Angelo G, Raposo G (2018) Shedding light on the cell biology of extracellular vesicles. Nat Rev Mol Cell Biol 19:213–228
20. Akuma P, Okagu OD, Udenigwe CC (2019) Naturally occurring exosome vesicles as potential delivery vehicle for bioactive compounds. Front Sustain Food Syst 3:23
21. He C, Zheng S, Luo Y, Wang B (2018) Exosome theranostics: biology and translational medicine. Theranostics 8:237–255
22. Aryani A, Denecke B (2016) Exosomes as a nanodelivery system: a key to the future of neuromedicine? Mol Neurobiol 53:818–834
23. Pascucci L, Coccè V, Bonomi A et al (2014) Paclitaxel is incorporated by mesenchymal stromal cells and released in exosomes that inhibit in vitro tumor growth: a new approach for drug delivery. J Control Release 192:262–270
24. Gomari H, Forouzandeh Moghadam M, Soleimani M (2018) Targeted cancer therapy using engineered exosome as a natural drug delivery vehicle. Onco Targets Ther 11:5753–5762
25. Kalani A, Chaturvedi P (2017) Curcumin-primed and curcumin-loaded exosomes: potential neural therapy. Neural Regen Res 12:205–206
26. Haney MJ, Klyachko NL, Zhao Y et al (2015) Exosomes as drug delivery vehicles for Parkinson's disease therapy. J Control Release 207:18–30
27. Faruqu FN, Xu L, Al-Jamal KT (2018) Preparation of exosomes for siRNA delivery to cancer cells. J Vis Exp. https://doi.org/10.3791/58814
28. Beuzelin D, Kaeffer B (2018) Exosomes and miRNA-loaded biomimetic nanovehicles, a focus on their potentials preventing type-2-diabetes linked to metabolic syndrome. Front Immunol 9:2711
29. Goh WJ, Zou S, Ong WY, Torta F, Alexandra AF, Schiffelers RM, Storm G, Wang J-W, Czarny B, Pastorin G (2017) Bioinspired cell-derived nanovesicles versus exosomes as drug delivery systems: a cost-effective alternative. Sci Rep 7:14322
30. Théry C, Amigorena S, Raposo G, Clayton A (2006) Isolation and characterization of exosomes from cell culture supernatants and biological fluids. Curr Protoc Cell Biol 30:3.22.1–3.22.29
31. Yu L-L, Zhu J, Liu J-X, Jiang F, Ni W-K, Qu L-S, Ni R-Z, Lu C-H, Xiao M-B (2018) A comparison of traditional and novel methods for the separation of exosomes from human samples. Biomed Res Int 2018:1–9
32. Patel GK, Khan MA, Zubair H, Srivastava SK, Khushman M, Singh S, Singh AP (2019) Comparative analysis of exosome isolation methods using culture supernatant for optimum yield, purity and downstream applications. Sci Rep 9:5335
33. Konoshenko MY, Lekchnov EA, Vlassov AV, Laktionov PP (2018) Isolation of extracellular vesicles: general methodologies and latest trends. Biomed Res Int 2018:1–27
34. Wang J-W, Zhang Y-N, Sze S et al (2017) Lowering low-density lipoprotein particles in plasma using dextran sulphate co-precipitates procoagulant extracellular vesicles. Int J Mol Sci 19:94
35. Jo W, Jeong D, Kim J, Cho S, Jang SC, Han C, Kang JY, Gho YS, Park J (2014) Microfluidic fabrication of cell-derived nanovesicles as endogenous RNA carriers. Lab Chip 14:1261–1269
36. Yoon J, Jo W, Jeong D, Kim J, Jeong H, Park J (2015) Generation of nanovesicles with sliced cellular membrane fragments for exogenous material delivery. Biomaterials 59:12–20
37. Jang SC, Kim OY, Yoon CM, Choi D-S, Roh T-Y, Park J, Nilsson J, Lötvall J, Kim Y-K, Gho YS (2013) Bioinspired exosome-mimetic nanovesicles for targeted delivery of chemotherapeutics to malignant tumors. ACS Nano 7:7698–7710
38. Jo W, Kim J, Yoon J, Jeong D, Cho S, Jeong H, Yoon YJ, Kim SC, Gho YS, Park J (2014)

Large-scale generation of cell-derived nanovesicles. Nanoscale 6:12056–12064

39. Grossman JG, Nywening TM, Belt B, Ahlers M, Hawkins WG, Strasberg SM, Goedegebuure PS, Linehan D, Fields RC (2017) Targeting inflammatory monocytes in human metastatic colorectal cancer. J Clin Oncol 35:605–605
40. Goh WJ, Lee CK, Zou S, Woon EC, Czarny B, Pastorin G (2017) Doxorubicin-loaded cell-derived nanovesicles: an alternative targeted approach for anti-tumor therapy. Int J Nanomed 12:2759–2767
41. Goh WJ, Zou S, Lee CK, Ou Y-H, Wang J-W, Czarny B, Pastorin G (2018) EXOPLEXs: chimeric drug delivery platform from the fusion of cell-derived nanovesicles and liposomes. Biomacromolecules 19:22–30
42. Fritze A, Hens F, Kimpfler A, Schubert R, Peschka-Süss R (2006) Remote loading of doxorubicin into liposomes driven by a transmembrane phosphate gradient. Biochim Biophys Acta Biomembr 1758:1633–1640
43. Sun D, Zhuang X, Xiang X, Liu Y, Zhang S, Liu C, Barnes S, Grizzle W, Miller D, Zhang H-G (2010) A novel nanoparticle drug delivery system: the anti-inflammatory activity of curcumin is enhanced when encapsulated in exosomes. Mol Ther 18:1606–1614
44. Takahashi Y, Nishikawa M, Shinotsuka H, Matsui Y, Ohara S, Imai T, Takakura Y (2013) Visualization and in vivo tracking of the exosomes of murine melanoma B16-BL6 cells in mice after intravenous injection. J Biotechnol 165:77–84
45. Smyth T, Kullberg M, Malik N, Smith-Jones P, Graner MW, Anchordoquy TJ (2015) Biodistribution and delivery efficiency of unmodified tumor-derived exosomes. J Control Release 199:145–155
46. Abe M, Havre PA, Urasaki Y, Ohnuma K, Morimoto C, Dang LH, Dang NH (2011) Mechanisms of confluence-dependent expression of CD26 in colon cancer cell lines. BMC Cancer 11:51
47. Ruutu M, Johansson B, Grenman R, Syrjänen K, Syrjänen S (2004) Effect of confluence state and passaging on global cancer gene expression pattern in oral carcinoma cell lines. Anticancer Res 24:2627–2631
48. Saeki K, Yuo A, Kato M, Miyazono K, Yazaki Y, Takaku F (1997) Cell density-dependent apoptosis in HL-60 cells, which is mediated by an unknown soluble factor, is inhibited by transforming growth factor beta1 and overexpression of Bcl-2. J Biol Chem 272:20003–20010
49. Fang C-Y, Wu C-C, Fang C-L, Chen W-Y, Chen C-L (2017) Long-term growth comparison studies of FBS and FBS alternatives in six head and neck cell lines. PLoS One. https://doi.org/10.1371/journal.pone.0178960
50. Shelke GV, Lässer C, Gho YS, Lötvall J (2014) Importance of exosome depletion protocols to eliminate functional and RNA-containing extracellular vesicles from fetal bovine serum. J Extracell Vesicles 3:24783
51. Li P, Kaslan M, Lee SH, Yao J, Gao Z (2017) Progress in exosome isolation techniques. Theranostics 7:789–804
52. Witwer KW, Buzás EI, Bemis LT et al (2013) Standardization of sample collection, isolation and analysis methods in extracellular vesicle research. J Extracell Vesicles. https://doi.org/10.3402/jev.v2i0.20360
53. Rhodes DG, Laue TM (2009) Determination of protein purity. In: Methods enzymol. Academic Press, New York, pp 677–689
54. Lu GW, Gao P (2010) Emulsions and microemulsions for topical and transdermal drug delivery. In: Personal Care & Cosmetic Technology, Handbook of Non-invasive Drug Delivery Systems, VS. Kulkarni (ed), William Andrew Publishing, pp 59–94, ISBN 9780815520252, https://doi.org/10.1016/B978-0-8155-2025-2.10003-4. (http://www.sciencedirect.com/science/article/pii/B9780815520252100034)
55. Gumustas M, Sengel-Turk CT, Gumustas A, Ozkan SA, Uslu B (2017) Effect of polymer-based nanoparticles on the assay of antimicrobial drug delivery systems. In: Multifunctional systems for combined delivery, biosensing and diagnostics (pp 67–108)
56. Malvern Instruments (2009) Technical Manual: Zetasizer Nano
57. Lobb RJ, Becker M, Wen SW, Wong CSF, Wiegmans AP, Leimgruber A, Möller A (2015) Optimized exosome isolation protocol for cell culture supernatant and human plasma. J Extracell Vesicles 4:27031
58. Zhu L, Gangadaran P, Kalimuthu S, Oh JM, Baek SH, Jeong SY, Lee S-W, Lee J, Ahn B-C (2018) Novel alternatives to extracellular vesicle-based immunotherapy—exosome mimetics derived from natural killer cells. Artif Cells Nanomed Biotechnol 46:S166–S179
59. Emam SE, Ando H, Selim A et al (2018) A novel strategy to increase the yield of exosomes (extracellular vesicles) for an expansion of basic research. Biol Pharm Bull 41:733–742
60. Fuhrmann G, Serio A, Mazo M, Nair R, Stevens MM (2015) Active loading into extracellular vesicles significantly improves the cellular

uptake and photodynamic effect of porphyrins. J Control Release 205:35–44

61. Lu M, Zhao X, Xing H, Xun Z, Zhu S, Lang L, Yang T, Cai C, Wang D, Ding P (2018) Comparison of exosome-mimicking liposomes with conventional liposomes for intracellular delivery of siRNA. Int J Pharm 550:100–113
62. Kaneti L, Bronshtein T, Malkah Dayan N, Kovregina I, Letko Khait N, Lupu-Haber Y, Fliman M, Schoen BW, Kaneti G, Machluf M (2016) Nanoghosts as a novel natural nonviral gene delivery platform safely targeting multiple cancers. Nano Lett 16:1574–1582
63. Krishnamurthy S, Gnanasammandhan MK, Xie C, Huang K, Cui MY, Chan JM (2016) Monocyte cell membrane-derived nanoghosts for targeted cancer therapy. Nanoscale 8:6981–6985
64. Lupu-Haber Y, Bronshtein T, Shalom-Luxenburg H et al (2019) Pretreating mesenchymal stem cells with cancer conditioned-media or proinflammatory cytokines changes the tumor and immune targeting by nanoghosts derived from these cells. Adv Healthc Mater 8:1801589. https://doi.org/10.1002/adhm.201801589
65. Bronshtein T, Toledano N, Danino D, Pollack S, Machluf M (2011) Cell derived liposomes expressing CCR5 as a new targeted drug-delivery system for HIV infected cells. J Control Release 151:139–148
66. Nakase I, Futaki S (2015) Combined treatment with a pH-sensitive fusogenic peptide and cationic lipids achieves enhanced cytosolic delivery of exosomes. Sci Rep 5:10112
67. Lin Y, Wu J, Gu W, Huang Y, Tong Z, Huang L, Tan J (2018) Exosome-liposome hybrid nanoparticles deliver CRISPR/Cas9 system in MSCs. Adv Sci (Weinheim, Baden-Wurttemberg, Germany) 5:1700611
68. Sato YT, Umezaki K, Sawada S, Mukai S, Sasaki Y, Harada N, Shiku H, Akiyoshi K (2016) Engineering hybrid exosomes by membrane fusion with liposomes. Sci Rep 6:21933
69. Piffoux M, Silva AKA, Wilhelm C, Gazeau F, Tareste D (2018) Modification of extracellular vesicles by fusion with liposomes for the design of personalized biogenic drug delivery systems. ACS Nano 12:6830–6842
70. Kooijmans SAA, Fliervoet LAL, van der MR, MHAM F, HFG H, en HPMP v B, Vader P, Schiffelers RM (2016) PEGylated and targeted extracellular vesicles display enhanced cell specificity and circulation time. J Control Release 224:77–85
71. Goh WJ, Zou S, Czarny B, Pastorin G (2018) nCVTs: a hybrid smart tumour targeting platform. Nanoscale 10:6812–6819
72. Illes B, Hirschle P, Barnert S, Cauda V, Wuttke S, Engelke H (2017) Exosome-coated metal–organic framework nanoparticles: an efficient drug delivery platform. Chem Mater 29:8042–8046
73. Fang RH, Hu C-MJ, Luk BT, Gao W, Copp JA, Tai Y, O'Connor DE, Zhang L (2014) Cancer cell membrane-coated nanoparticles for anticancer vaccination and drug delivery. Nano Lett 14:2181–2188
74. Jin J, Krishnamachary B, Barnett JD, Chatterjee S, Chang D, Mironchik Y, Wildes F, Jaffee EM, Nimmagadda S, Bhujwalla ZM (2019) Human cancer cell membrane-coated biomimetic nanoparticles reduce fibroblast-mediated invasion and metastasis and induce T-cells. ACS Appl Mater Interfaces 11:7850–7861

Chapter 12

Hydrogel Beads of Natural Polymers as a Potential Vehicle for Colon-Targeted Drug Delivery

Janarthanan Pushpamalar, Thenapakiam Sathasivam, and Michelle Claire Gugler

Abstract

Polysaccharides are excellent candidates for drug delivery applications as they are available in abundance from natural sources. Polysaccharides such as starch, cellulose, lignin, chitosan, alginate, and tragacanth gum are used to make hydrogels beads. Hydrogels beads are three-dimensional, cross-linked networks of hydrophilic polymers formed in spherical shape and sized in the range of 0.5–1.0 mm of diameter. Beads are formed by various cross-linking methods such as chemical and irradiation methods. Natural polymer-based hydrogels are biocompatible and biodegradable and have inherently low immunogenicity, which makes them suitable for physiological drug delivery approaches. The cross-linked polysaccharide-based hydrogels are environment-sensitive polymers that can potentially be used for the development of "smart" delivery systems, which are capable of control release of the encapsulated drug at a targeted colon site. This topic focuses on various aspects of fabricating and optimizing the cross-linking of polysaccharides, either by a single polysaccharide or mixtures and also natural-synthetic hybrids to produce polymer-based hydrogel vehicles for colon-targeted drug delivery.

Key words Natural polymers, Beads, Drug delivery system, Colonic system

1 Introduction

A drug delivery system is defined as a method or process of introducing and delivering the therapeutic agents into the body and achieve a therapeutic effect [1]. Conventional drug delivery system usually results in a wide range of fluctuation in drug concentration in the bloodstream and immediate release of drug which the amount and the rate of drug release is difficult to be controlled [2]. This system will also cause low patient compliance and will require frequent administration. In order to overcome these problems, sustained drug delivery system has been introduced to give a breakthrough for drug delivery in the pharmaceutical technology field.

Kumaran Narayanan (ed.), *Bio-Carrier Vectors: Methods and Protocols*, Methods in Molecular Biology, vol. 2211, https://doi.org/10.1007/978-1-0716-0943-9_12,

The colon is an area where both the local and systemic delivery of drugs can take place. Colon-targeted drug delivery aims to limit the release and absorption of an orally administered drug to the colon to locally treat associated diseases, including irritable bowel syndrome or Crohn's disease. For successfully delivering the drug to the colonic area (the targeted site), the drug has to be masked to protect from microbial and pH gradient degradations and release of the drug while passing through the stomach and intestine before reaching the colon [3].

Drug delivery approaches using biodegradable polysaccharide hydrogel beads as drug carriers have shown a prospective to potentiate the therapeutic efficiency of orally administered drugs by allowing a decreased dose to achieve desired therapeutic effects [4]. This advantage ultimately addresses the challenges related to systemic side effects that are currently faced through conventionally oral drug administration. Hydrogels possess several desirable properties that make them an excellent drug delivery vehicle. They provide a three-dimensional network with a high water content that can encapsulate hydrophilic drugs (water-soluble) and prevent premature enzymatic degradation by obstructing the diffusion of enzymes into the inner space of the spherical hydrogel bead [5]. Additionally, the adjustable physical and degradation elements of hydrogel beads enable control over the drug-releasing properties under physiological conditions [6].

These potential drug carriers allow an improvement of drug stability as well as efficacy and reduction of adverse drug effects by enabling a sustained and controlled drug release locally within the colon. The drug-releasing properties are subjected to the bead's polysaccharide formulation and drug loading efficiency. Therefore, the release kinetics of the drugs from the polysaccharide beads are limited to the conditions found in the colonic environment externally acting on the beads [7].

It is crucial that pH-sensitive polymers are used to fabricate the beads so they can withstand different levels of pH at varying transit times to only initiate their swelling dissolution in colon site-specific pH. As shown in Fig. 1, considering that the orally administered bead needs to travel through pH-varying environments in the gastrointestinal tract, including an acidic stomach of pH 1–2, an almost neutral small intestine with a pH ranging from pH 6.5–7.5, and ultimately to a colon with a pH of 5.7–7.0 [4].

In fabricating polymer-based beads as drug delivery vehicles, the choice of pH-sensitive polymer is the first important aspect. The most cost-efficient, structurally degradable polymer for the intended application should be considered. Some biodegradable polymers require functionalization in order to adjust their hydrophilicity and resulting swelling properties to a maximum under the conditions of the targeted site of drug delivery.

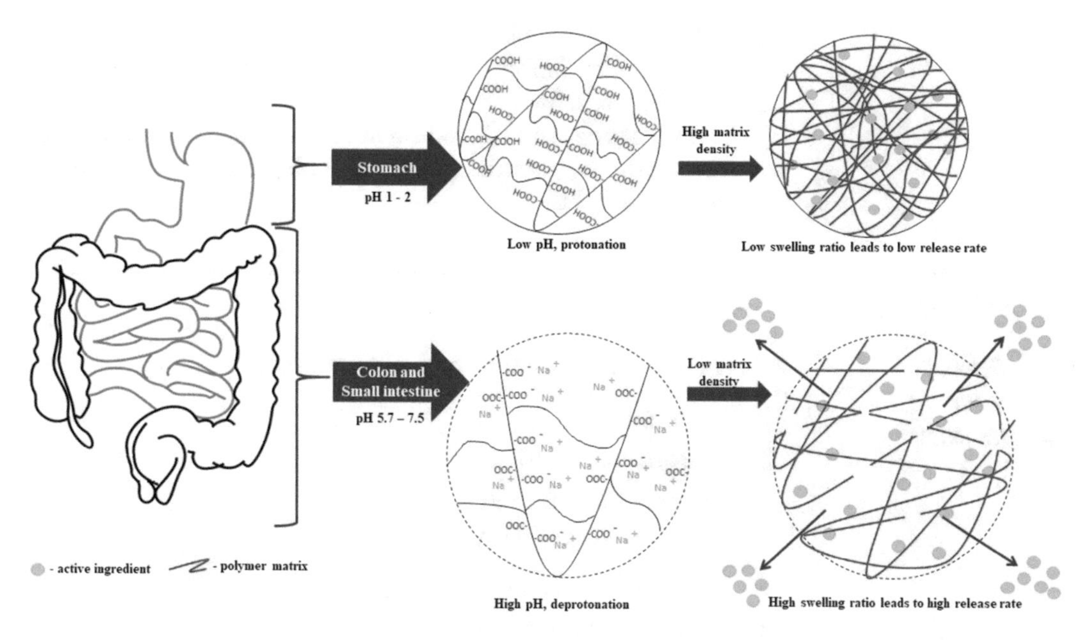

Fig. 1 The swelling behavior of pH-sensitive hydrogel beads in the gastrointestinal tract, at different pHs in the stomach, intestine, and colon

In the fabrication process and formulation of the polymer beads, the polysaccharide mixture, the cross-linking approaches of the drug-carrying beads, and the resulting cross-linking densities inaugurate altering drug release profiles [5]. An increased beads matrix density by increasing the cross-linking density will lead to a slower release of the drug out of the bead.

The chemical cross-linking approaches provides a stronger and less flexible polymer matrix but is not necessarily reproducible. However, radiation cross-linking using gamma, UV radiation, or electron beams gives a replicable breakage and reformation of chemical bonds between the polymer molecules [5]. The cross-linking approach of complex coacervation depends on electrostatic attraction when two oppositely charged polymers are mixed, and it is a low-cost technique that ensures mechanical and thermal protection of the bead encapsulated drug [5]. However, aggregation and clumping are common limitations in the polymer beads recovery.

Lastly, ionotropic gelation is another method that cross-links and induces gelation of polyelectrolyte polymers in the presence of counterions. It can be done in a conventional approach where the polyelectrolyte polymer solution is manually dropped into the cross-linking solution [7]. In order to prepare a homogenous sized beads, a Buchi encapsulator is used for a more consistent ionotropic gelation [8, 9]. Natural polysaccharide beads have been produced via ionotropic gelation using chemical cross-linkers, including aluminum chloride and calcium chloride, complex coacervation, and irradiation cross-linking or combinations of above-mentioned approaches.

In this chapter, the fabrication aspects of polysaccharide-based hydrogel beads as potential carriers for colon-targeted drug delivery will be discussed.

2 Materials

2.1 General Chemicals and Reagents

All reagents used should be of analytical grade.

1. Methanol.
2. Isopropanol (99.9%).
3. Ethanol (denatured 95%).
4. Sodium chlorite (80%).
5. Sodium monochloroacetate (98%).
6. Sodium hydroxide pellets (97%).
7. Glacial acetic acid (99.9%).
8. Acetone (99.6%).
9. Chloroform.
10. Formic acid.
11. Hydrogen peroxide (30%).
12. Chloroform.
13. Sulfuric acid (98.7%).
14. Formic acid.
15. Phenolphthalein.
16. 0.9% Saline: Dissolve 0.9 g NaCl in 70 mL deionized water, and bring up to 100 mL.
17. Hielscher UIP500hd ultrasonic homogenizer.

2.2 Materials and Equipment

1. Automated 3 × 1 systematic multi hot plate stirrer (e.g., WiseStir, model SMHS-3, Korea).
2. Buchi encapsulator that meets the following technical specifications: (2 mm nozzle, 10 cm distance between nozzle and solution, flow rate of 0.8 mL/min, electrode voltage of 900 mV, vibration frequency of 1700 Hz, temperature: 35 °C [8]; 750 μm nozzle, air pressure of 500–700 mbar, electrode voltage of 1000 mV, vibration frequency of 600 Hz [9].
3. 3 mL (diameter 8.95 mm) and 5 mL (diameter 13 mm) plastic syringes.
4. 18- and 21-gauge needle tips.

5. Atomizer spray bottle.
6. Hair dryer.

2.3 Aluminium and Irradiation Cross-Linked Beads by Conventional Ionotropic Gelation

1. Polymers: Carboxymethyl sago pulp (CMSP) from sago waste (PPES Sago Industries (Mukah) Sdn. Bhd, Sarawak, Malaysia).
2. Cross-linkers: Aluminum chloride hexahydrate, electron beam irradiation.
3. Active ingredient: 5-aminosalicylic acid (5-ASA).
4. 10% (w/v) CMSP solution (*see* **Note 1**): Prepare using 50.0 mL distilled water with pH adjustment using acetic acid to pH 4 (*see* **Note 2**).

2.4 Calcium and Irradiation Cross-Linked Beads by Conventional Ionotropic Gelation

1. Polymers: Carboxymethyl sago pulp (CMSP) from sago waste was obtained from Ng Kia Heng Kilang Sagu Industries, Batu Pahat, Johor, pectin (degree of esterification 63–66%).
2. Cross-linker: Anhydrous calcium chloride.
3. Active ingredient: Diclofenac sodium.

2.5 Iron Cross-Linked Beads by Complex Coacervation and Conventional Ionotropic Gelation

1. Polymers: Carboxymethylcellulose (degree of substitution: 0.6–0.9 and viscosity of 400–1000 mPa s, 2% in H_2O at 25 °C).
2. Cross-linker: Ferric chloride.
3. Active ingredient: Ibuprofen USP 32.

2.6 Aluminium Cross-Linked Beads by Emulsification and Vibration Technology of Ionotropic Gelation

1. Polymers: Carboxymethyl sago cellulose (CMSC) from sago waste obtained from Ng Kia Heng Kilang Sago Industries, Batu Pahat, Johor, Malaysia, polyethylene glycol (PEG), glycerine.
2. Cross-linker: Aluminium choride hexahydrate.
3. Active ingredient: Red palm oil from Carotino Premium, Malaysia.

2.7 Calcium Cross-Linked by Vibration Technology of Ionotropic Gelation and Surface Coated Beads

1. Polymers: Sodium alginate (brown algae), Ispaghula (*Plantago ovata*), Zein (maize).
2. Cross-linker: Calcium chloride.
3. Active ingredient: Diclofenac sodium.
4. 10% (w/v) zein solution: Prepared in 80% (v/v) ethanol.

2.8 Aluminium Cross-Linked Beads by Conventional Ionotropic Gelation

1. Polymers: Carboxymethyl tragacanth (CM-TG) from tragacanth gum.
2. Cross-linker: Aluminum chloride hexahydrate.
3. Active ingredient: Diclofenac sodium.

4. 10/20% (w/v) CM-TG: Weigh 10/20 g of CM-TG of DS 1.2 into a volumetric flask (100 mL) and add distilled water up to the graduated mark. Stir the mixture at 600 rpm for 10 min using an automated stirrer. Adjust the pH of the mixture to 4 using glacial acetic acid.

3 Methods

3.1 Aluminium and Irradiation Cross-linked Beads by Conventional Ionotropic Gelation

1. Incorporate 5-ASA into 50 mL of 10% CMSP solution by adding 10, 20, and 30% (w/w) of 5-ASA of total CMSP used directly and using automated magnetic stirrer for 15 min at 800 rpm (*see* **Note 3**). Add 0.1% (w/w) of sodium metabisulfite, sonicate at 80% amplitude for 2 min [7].
2. Use the conventional ionotropic method by dripping the 10 mL of CMSP/5-ASA solution into a stirring 5% (w/v) aluminum chloride ($AlCl_3$) solution (*see* **Note 4**) that contains 0.1% (w/w) sodium metabisulfite using a 10 mL syringe with 18-gauge needle tip at a distance of 6 cm from the cross-linking solution (*see* **Note 5**) within a 2 min introduction time span to form the beads. Leave the beads to cure for 15 min, then collect and dry them at 40 °C for 24 h to constant weight. Store the beads at 10 °C.
3. For further cross-linking, subject freshly prepared CMSP/5-ASA beads to electron beam irradiation at 10 kGy (*see* **Notes 6** and 7). Then, dry the beads at 40 °C for 24 h to constant weight. Store the beads at 10 °C.

3.2 Calcium and Irradiation Cross-Linked Beads by Conventional Ionotropic Gelation

1. Dissolve 10% (w/w) of diclofenac sodium based on the CMSP and pectin total weight in distilled water. Dissolve 15%/5% (w/v) CMSP/pectin in the same solution stir for 30 min at 600 rpm using automated magnetic stirrer followed by homogenization (*see* **Note 8**) [10].
2. Use the conventional ionotropic gelation method within a 2 min introduction time span to form the beads by manually dripping 10 mL of CMSP/pectin/diclofenac solution into a magnetically stirring 10% (w/v) calcium chloride solution at pH 3.6 (*see* **Note 9**). Use a 10 mL syringe with a 21-gauge hypodermic needle tip at a distance of 6 cm from the cross-linking solution (*see* **Note 5**). Leave the beads to cure for 20 min. Remove the beads via filtration and wash using distilled water.
3. For further cross-linking, subject the beads to electron beam irradiation at 5, 10, 15, and 20 kGy (*see* **Note 10**). Dry at 37 °C for 24 h to obtain constant weight.

3.3 Iron Cross-Linked Beads by Complex Coacervation and Conventional Ionotropic Gelation

1. Dissolve 4 g of CMC and gelatine separately in distilled water (45 °C). Adjust the gelatine solution to pH 7, add 6 g of ibuprofen (*see* **Note 11**) and stir at 40 °C for 30 min. Add the CMC solution (*see* **Note 12**) [11].
2. Use the complex coacervation method to prepare the polymer-drug beads by dropping 10% (w/v) HCl into a polymer-drug solution, cooling it to 5 °C. Form microcapsules by dropping the polymer-drug coacervate solution into 50 mL of 30% (w/v) aqueous ferric chloride (*see* **Note 13**).
3. Drop the microcapsules into 15% (w/v) ferric chloride in a hydro-alcoholic solvent using an 18-gauge needle attached to a 3 mL syringe (*see* **Note 14**). Leave the microcapsules to cross-link for 1 h at 5 °C.
4. Collect the beads using a nylon mesh and wash once using 100 mL distilled water and twice using 100 mL of 2-propanol. Dry at 40 °C for 6 h.

3.4 Aluminium Cross-Linked Beads by Emulsification and Vibration Technology of Ionotropic Gelation

1. Dissolve 6.25 g of CMSC in 50 mL degassed distilled water and add 30% (w/v) glycine and polyethylene glycol as a plasticizer (*see* **Note 15**). Using emulsification, add the red palm oil (*see* **Note 16**) gradually into the magnetically stirring CMSC solution and homogenize the resulting solution using automated magnetic stirrer at 10,000 rpm for 5 min (*see* **Note 17**) [8].
2. Apply the external ionotropic gelation method using a syringe pump. Set a Buchi encapsulator with a 2 mm nozzle positioned 10 cm above a 200 mL of a 5% (v/v) aluminum chloride solution (*see* **Note 18**). Adjust the settings to a flow rate of 0.8 mL/min, electrode voltage of 900 mV, vibration frequency of 1700 Hz, and 35 °C (*see* **Note 19**).
3. Leave the beads to cross-link for 1 h. Collect the beads using a plastic mesh and wash using distilled water. Freeze the oil loaded beads at −80 °C overnight and then freeze-dry for 24 h (*see* **Note 20**). Store the beads in a desiccator (*see* **Note 21**).

3.5 Calcium Cross-Linked by Vibration Technology of Ionotropic Gelation and Surface Coated Beads

1. Dissolve 5 g of sodium alginate in 250 mL distilled water using a magnetic stirrer. Stir 500 mg of ispaghula husk overnight in 100 mL of distilled water and remove debris using a muslin cloth (*see* **Note 22**). Add the ispaghula filtrate into the alginate solution [9].
2. Triturate 25–30 mL of the ispaghula-alginate solution onto ground 5 g of ibuprofen (*see* **Note 11**). Stir the ibuprofen-alginate solution and add the remaining alginate solution.

3. Immediately use the external ionotropic gelation method via a Buchi encapsulator with a 750 μm nozzle positioned above a 5% (w/v) calcium chloride solution. Adjust the settings to an air pressure of 500–700 mbar, electrode voltage of 1000 mV, and vibration frequency of 600 Hz (*see* **Note 19**). Leave the beads to cure for 30 min. Remove the beads using a nylon mesh and wash using distilled water repeatedly. Dry the beads at 40 °C until constant weight and store them in glass vials at room temperature.
4. To coat the beads with zein, a manual atomizer spray bottle is used to spray 5 mL of 10% zein solution onto 1 g of uniform layered polymer-drug beads (*see* **Note 23**). Dry the beads using a conventional hair dryer at 40–50 °C. Repeat the same spraying process after turning the beads to ensure an even coating.

3.6 Aluminium Cross-Linked Beads by Conventional Ionotropic Gelation

1. Prepare the CM-TG diclofenac beads by dissolving 100 mL 10/20% (w/v) of CM-TG (DS 1.2) and adding 10/20% (w/v) of sodium diclofenac (*see* **Note 24**), respectively. Stir using automated magnetic stirrer at 600 rpm and subject the solution to sonication for 1 min using Hielscher UIP500hd ultrasonic homogenizer at 60% amplitude [12].
2. Use the conventional ionotropic method by dropping the CM-TG diclofenac solution into a magnetically stirred 5% (w/v) aluminum chloride solution using a 10 mL syringe with 21-gauge needle positioned at a 6 cm distance from the cross-linking solution. Leave the beads to cross-link for 15 min (*see* **Note 25**).

4 Notes

1. 10% (w/v) CMSP solution was chosen as any concentration range below 10% (w/v) (2.5–7.5%) would not provide sufficient viscosity to yield stable beads. A CMSP concentration above 10% (w/v) would cause excessive viscosity resulting in extruding difficulties of the CMSP solution using the conventional method. This yields the formation of teardrop-shaped beads. Due to elevated viscosity, the formation of air bubbles could be induced (in reference to Fig. 2).
2. The CMSC solution is adjusted to pH 4.3 to allow 5-ASA to be predominantly present as insoluble dipolar zwitterionic species by approaching its isoelectric point. This leads to a higher drug loading efficiency as the degree of solubility promotes the loss of the drug of the polymer matrix into the gelation medium during the ionotropic gelation process. The stability of the 5-ASA was tested in different pH media and found that it is stable up to pH 7 (in reference to Fig. 3).

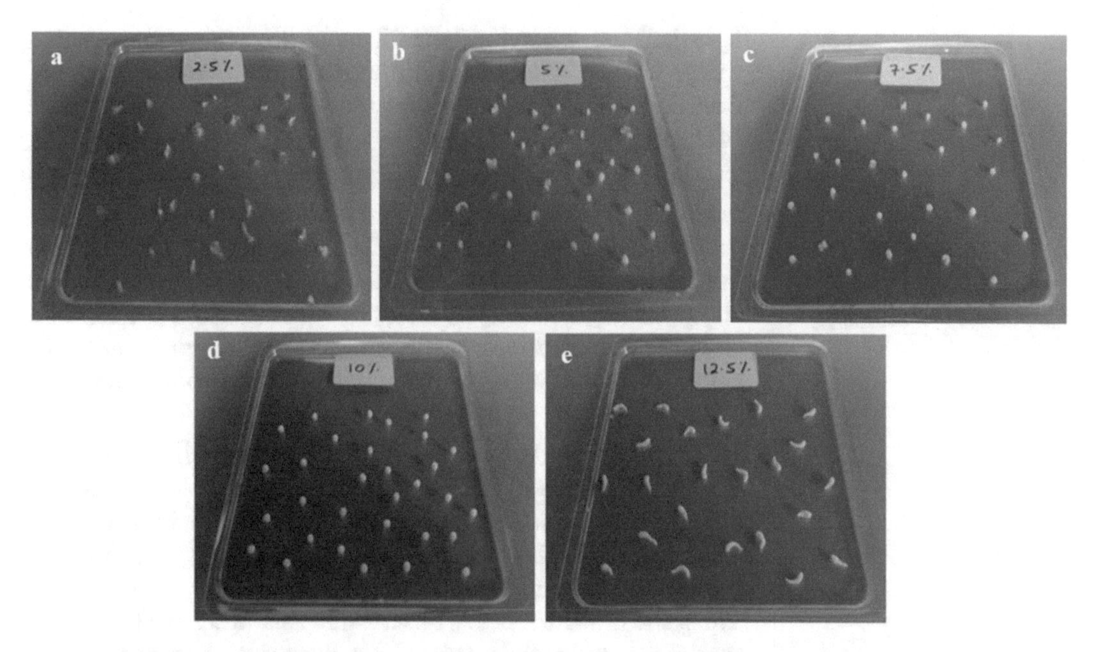

Fig. 2 The hydrogel beads of teardrop-shaped formed using various carboxymethylcellulose (CMC) concentrations by ionotropic cross-linking

Fig. 3 The stability of the 5-ASA in different pH media

3. A 5-ASA concentration exceeding 30% causes the formation of flattened beads that adhere strongly to the surface of the drying plates.
4. A 5% $AlCl_3$ cross-linker solution was used since elevated concentrations cause too rapid cross-linking, leaving the beads matrix too compact, resulting in a drug loading efficiency reduction.
5. The limiting factor to the formation of uniform and spherically shaped bead is defined by the inability to control the flow rate and pressure parameters exerted by the manual ionotropic gelation. The flow rate of the drop addition can be regulated

by using a stopwatch. The individual drops should be added within a set of introduction time to minimize bead agglomeration. Sodium metabisulfite was added to reduce the adherence of the beads to one another.

6. The irradiation dose (kGy) was chosen at 10 kGy since higher radiation dose would cause more scission and oxidative degradation leading to a dilution of the viscous CMSP solution. Furthermore, the active ingredient 5-ASA undergoes thermal and radio degradation, reducing its effectiveness when exposed to higher irradiation doses. Lower kGy doses would not provide sufficient cross-linking rather than predomination of scission.
7. Higher drug loadings could distort the bead shape, possibly due to disruption of the matrix by the drug molecules during bead formation and also by exhibiting more resistance when extruding through the needle. The average diameter of 50–100 beads should be measured to optimize the size of the beads. Thus, the unloaded beads observed in this study will be smaller than the drug-loaded beads. The average size and size deviation will increase with increasing 5-ASA loading. It could be due to an increase in viscosity because of higher drug concentration in the polymeric mixture.
8. The pectin proportion cannot exceed 5% (w/v) as its gelling properties would cause the solution to be excessively viscous.
9. The solubility of diclofenac sodium is dependent on the pH of the dissolution medium. It is soluble at pH 5–8, whereas an acidified dissolution medium would result in precipitation of the diclofenac salt.
10. Subjected to electron beam irradiation to cross-link for the second time to tighten the matrix by increasing the extent of the cross-linking and hold the drug molecules tightly in the polymer matrix.
11. Micronization of ibuprofen would improve the encapsulation efficiency by increasing the surface area of the solid drug particles.
12. Measuring the electrophoretic mobility of the gelatine and carboxymethylcellulose as a function of pH and ionic strength would ensure that the optimum conditions for the highest degree of coacervation would be met. Approaching zero mobility charge would prevent the presence of excessive polymers in the coacervate medium.
13. Ferric chloride is toxic for human consumption; thus, residual presence on the sample should be investigated using gas chromatography.

14. 18-gauge needle size was used to prevent clogging of the coacervate complex extrusion.
15. Plasticizer addition affects the oil loading of the bead but improves its fragile characteristics to a smoother surface and shape. Adding PEG-400 improves the thermal stability of the beads.
16. The beads size should not change regardless of the oil loading concentration. The loading efficiency is independent of the oil loading when it exceeds 40% since oil can leak through the polymer matrix into the cross-linking solution due to phase separation.
17. Emulsification allows a temporarily stable mixture of immiscible polymer and oil. Elevated shear flow allows a homogenizer to disperse the immiscible oil into small droplets, which cannot be done by stirring.
18. A 5% $AlCl_3$ cross-linker solution is used since elevated concentrations (e.g., 10%) would cause too rapid cross-linking that compacts the beads matrix excessively, leading to the oil leakage.
19. Using vibration technology-based ionotropic gelation allows the liquid jet to be prilled into small droplets by using vibrational frequencies. This enables the reproducible production of mono-dispersed and homogeneous particles.
20. Conventional oven drying flattens out the shape of the bead. Freeze drying is advantageous as it minimizes oxidation of oil by not exposing the beads to heat.
21. The percentage of oil lost during the storage phase should be investigated due to the porous nature of the polymer matrix allowing the oil to seep out.
22. Ispaghula concentrations varies every time undissolved husk material was removed using muslin (cheese) cloth. There is no control over the ratio of dissolved and undissolved husk. The filtration is time consuming due to the viscous, non-flowing nature of ispaghula.
23. Using a conventional spray bottle to coat the beads with zein has minimal control over the thickness and even distribution of the coating material. It would be advised to use a pharmaceutical coating pan.
24. Drug entrapment efficiency is considered low with less than 40% compared to [7, 10, 11] despite increasing drug concentration to 20%. This is because tragacanth gum has a less rigid matrix, allowing the drug to diffuse through the matrix easily during cross-linking. The number of carboxymethylation of the tragacanth gum backbone chain is low and therefore is expected to yield a less rigid hydrogel matrix.

25. Investigating the release at pH 1.2 could be conducted to mimic the drug release while passing the stomach en route to the colon to confirm the balanced release of the drug.

References

1. Tiwari G, Tiwari R, Sriwastawa B et al (2012) Drug delivery systems: an updated review. Int J Pharm Investig 2(1):2
2. Ratnaparkhi MP, Somvanshi FU, Pawar SA et al (2013) Colon targeted drug delivery system. Int J Pharm Rev Res 2(8):33–42
3. Jose S, Dhanya K, Cinu T et al (2009) Colon targeted drug delivery: different approaches. J Young Pharm 1(1):13
4. Amidon S, Brown JE, Dave VS (2015) Colon-targeted oral drug delivery systems: design trends and approaches. AAPS PharmSciTech 16(4):731–741
5. Pushpamalar J, Veeramachineni A, Owh C et al (2016) Biodegradable polysaccharides for controlled drug delivery. ChemPlusChem 81:504–514
6. Li J, Mooney DJ (2016) Designing hydrogels for controlled drug delivery. Nat Rev Mater 1 (12):16071
7. Thenapakiam S, Kumar DG, Pushpamalar J et al (2013) Aluminium and radiation cross-linked carboxymethyl sago pulp beads for colon targeted delivery. Carbohydr Polym 94 (1):356–363
8. Sathasivam T, Muniyandy S, Chuah LH et al (2018) Encapsulation of red palm oil in carboxymethyl sago cellulose beads by emulsification and vibration technology: physicochemical characterization and in vitro digestion. J Food Eng 231:10–21
9. Lyn Heng J, Teng J, Saravanan M et al (2018) Influence of Ispaghula and Zein coating on ibuprofen-loaded alginate beads prepared by vibration technology: physicochemical characterization and release studies. Sci Pharm 86 (2):24
10. Tan HL, Tan LS, Wong YY et al (2016) Dual crosslinked carboxymethyl sago pulp/pectin hydrogel beads as potential carrier for colon-targeted drug delivery. J Appl Polym 133 (19):43416
11. Huei GOS, Muniyandy S, Sathasivam T et al (2016) Iron cross-linked carboxymethyl cellulose–gelatin complex coacervate beads for sustained drug delivery. Chem Papers 70 (2):243–252
12. Veeramachineni AK, Sathasivam T, Paramasivam R et al (2019) Synthesis and characterization of a novel pH-sensitive aluminum crosslinked carboxymethyl tragacanth beads for extended and enteric drug delivery. J Polym Environ 27(7):1516–1528

Chapter 13

Scaffold-Based Delivery of CRISPR/Cas9 Ribonucleoproteins for Genome Editing

Wai Hon Chooi, Jiah Shin Chin, and Sing Yian Chew

Abstract

The simple and versatile CRISPR/Cas9 system is a promising strategy for genome editing in mammalian cells. Generally, the genome editing components, namely Cas9 protein and single-guide RNA (sgRNA), are delivered in the format of plasmids, mRNA, or ribonucleoprotein (RNP) complexes. In particular, non-viral approaches are desirable as they overcome the safety concerns posed by viral vectors. To control cell fate for tissue regeneration, scaffold-based delivery of genome editing components will offer a route for local delivery and provide possible synergistic effects with other factors such as topographical cues that are co-delivered by the same scaffold. In this chapter, we detail a simple method of surface modification to functionalize electrospun nanofibers with CRISPR/Cas9 RNP complexes. The mussel-inspired bio-adhesive coating will be used as it is a simple and effective method to immobilize biomolecules on the surface. Nanofibers will provide a biomimicking microenvironment and topographical cues to seeded cells. For evaluation, a model cell line with single copies of enhanced green fluorescent protein (U2OS.EGFP) will be used to validate the efficiency of gene disruption.

Key words Gene delivery, Cas9 protein, Ribonucleoprotein, Tissue engineering, Electrospinning

1 Introduction

Clustered regularly interspaced short palindromic repeats (CRISPR)/Cas9 system is an efficient and simple-to-implement genome editing technique [1, 2]. In CRISPR/Cas9-based genome editing, two main components are needed: Cas9 protein, which is an RNA-guided DNA endonuclease, and single-guide RNA (sgRNA), which binds to target sequence complementarily [1–3]. The Cas9-sgRNA complex binds to a target sequence and induces a double-stranded break (DSB) at a specific target site. When a template is included, the DSB may be repaired by Homolog-Directed Repair (HDR) by inserting the template along the break site. Without this template, the DSB can be repaired by Non-Homologous End Joining (NHEJ) pathways

Kumaran Narayanan (ed.), *Bio-Carrier Vectors: Methods and Protocols*, Methods in Molecular Biology, vol. 2211, https://doi.org/10.1007/978-1-0716-0943-9_13,

that lead to deletion and insertion at the site. Hence, the gene or targeted locus will be disrupted.

There are a few methods to deliver Cas9 and sgRNA effectively. Generally, Cas9 can be delivered in the format of plasmid, mRNA, or protein [4, 5]. Plasmid and mRNA will be transcribed and translated into Cas9 protein in the cell cytoplasm. Subsequently, the translated Cas9 proteins will be complexed with sgRNA. Comparatively, Cas9-sgRNA ribonucleoprotein (RNP) complexes are more efficient in transfection and have the least off-target effects [6]. Nonetheless, these Cas9-sgRNA RNP complexes still require efficient delivery vehicles into the cells to achieve genomic editing. Here, non-viral approaches are desirable as they overcome the safety concerns posed by viral vectors [6, 7].

Scaffolds are frequently used in tissue engineering and regenerative medicine approaches to support tissue regrowth and deliver topographical cues that can mimic the native tissue microenvironment. These scaffolds can also be modified to locally deliver cells and therapeutics, such as proteins, small molecules, and nucleic acids to the injured tissues. Therefore, the incorporation of CRISPR/Cas9 systems with scaffolds offers a route to modulate cell behavior on the scaffold. Furthermore, such scaffold-mediated CRISPR/Cas9 systems can provide possible synergistic effects from other factors that are co-delivered by the same scaffold.

Immobilizing proteins or nucleic acids on scaffolds have been intensively reported [8, 9]. One method of immobilizing these biomolecules involves surface modification of the scaffold. In this chapter, a simple method of surface modification to functionalize electrospun nanofibers with CRISPR/Cas9 complexes will be described. Nanofibers will provide a biomimicking microenvironment and topographical cues to seeded cells.

Here, the mussel-inspired bio-adhesive coating will be used as it is a simple and effective method to immobilize biomolecules on surfaces. Such an approach has been demonstrated to be effective in delivering nucleic acids in the form of siRNAs [10–12] and miRNAs [13–16], as well as Cas9-sgRNA RNP complexes [17]. As a proof of concept, a model cell line with single copies of enhanced green fluorescent protein (U2OS.EGFP) will be used to validate the efficiency of genome editing.

2 Materials

2.1 Preparation of Electrospun Fiber Scaffolds with Bio-adhesives

1. Polycaprolactone (PCL, M_w: 80 and 40 kDa).
2. 2,2,2-Trifluoroethanol (TFE).
3. 18 mm diameter glass coverslips.
4. Carbon tape.

5. 22G needle.
6. 1 mL syringe.
7. Electrospinning setup: Syringe pump, rotating wheel with speed control, wires, power source.
8. Silicone glue.
9. 70% ethanol.
10. Distilled water (sterilized by 0.02 μm syringe filter).
11. 3,4-Dihydroxy-l-phenylalanine (L-DOPA).
12. Buffer for polymerization of DOPA: 10 mM Bicine, 250 mM NaCl, HCl (to adjust pH), pH 8.5.
13. Laminin mouse protein.
14. 1× PBS.

2.2 Preparation of Complexes

1. sgRNA targeting EGFP: GAAATTAATACGACTCACTA TAGGTGAACCGCATCGAGCTGAA GTTTTAGAGCTA GAAATAGCAAGTTAAAATAAGGCTAGTCCGTTAT CAACTTGAAAAAGTGGCACCGAGTCGGTGCTTTTTT (*see* **Note 1**).
2. Cas9 protein: EnGen Cas9 NLS protein (NEB) (*see* **Note 2**).
3. Transfection reagent: Lipofectamine CRISPRMAX Cas9 Transfection Reagent (Invitrogen) (*see* **Note 3**).
4. Opti-MEM.

2.3 Cell Seeding/Culture

1. U2OS.EGFP cells.
2. Dulbecco's Modified Eagle's Medium (DMEM) with glutaMAX.
3. Fetal bovine serum (FBS).
4. Penicillin-Streptomycin (10,000 U/mL, PS).
5. Culture medium for U2OS.EGFP cells: DMEM with glutaMAX containing 10% FBS and 1% PS.
6. Trypsin-EDTA (0.05%).

2.4 Evaluation of EGFP Expression

1. 4% Paraformaldehyde (PFA): 4 g of paraformaldehyde in 100 mL of 1× PBS.
2. 0.1% Triton-X 100: 0.05 mL of Triton-X 100 in 50 mL of 1× PBS.
3. Rhodamine Phalloidin.
4. DAPI.
5. 1× PBS.
6. 0.6 U/mL Proteinase K: working stock in 1× Tris–EDTA buffer (10 mM Tris–HCl, 0.1 mM EDTA, pH 8.0).

7. DNA purification kit (*see* **Note 4**).
8. Genomic cleavage assay kit (*see* **Note 5**).
9. Forward and Reverse primers for the PCR step in genomic cleavage assay: GACGTAAACGGCCACAAGTT (Forward). GCGGATCTTGAAGTTCACCT (Reverse).
10. Gel electrophoresis setup.
11. Agarose.
12. 10× Tris/Acetate/EDTA (TAE) buffer: 400 mM Tris–acetate, 10 mM EDTA, pH 8.0 (*see* **Note 6**).
13. 10 mg/mL Ethidium bromide (Bio-Rad) (*see* **Note 7**).
14. 1 kb DNA ladder.

3 Methods

3.1 Electrospun Fiber Scaffold Preparation

1. Prepare electrospun fiber scaffolds by electrospinning [14, 17] (*see* **Note 8**).
2. Melt 1 g of PCL (40 kDa) at 60 °C for 30 min to form a PCL block in a mold (*see* **Note 9**). After this PCL block is formed, trim the block down to 10 mm by 5 mm by 10 mm before using a cryostat to section this block into small strips of PCL films of 20 μm thickness.
3. Dissolve PCL (80 kDa) in TFE (14% W/V) overnight.
4. Vortex the solution until homogenous.
5. Before electrospinning, adhere the small strips of PCL films (40 kDa) with 20 μm thickness onto glass coverslips (diameter ϕ 18 mm) to create a 10 mm × 10 mm area. Place these PCL films onto the coverslips to form a square configuration before melting them on a hotplate, at 50 °C for 15 s. Remove coverslips from hotplate once these films turn transparent.
6. Fix 8 coverslips on a rotating collector ($\phi = 15$ cm) with carbon tapes.
7. Electrospin the PCL fibers onto the coverslips that are fixed on a rotating collector (2400 rpm). Apply a flow rate of 1.0 mL/h and voltages of +8 kV and −4 kV to the needle tip and rotating collector, respectively. Fix the spinneret-to-collector distance at 21 cm. Allow electrospinning to occur for 6 min 30 s (100 μL of 14% W/V PCL solution electrospun) before removing the coverslips off the rotating collector.
8. Secure the electrospun PCL fibers onto the coverslips with silicone glue. Using a fine paintbrush, apply a small amount of silicone glue along the parameter of the square configuration formed by small strips of PCL films (refer to **step 5**). Allow

silicone glue to dry overnight at room temperature. Prior to surface modification, store scaffolds in 12-well plates for no more than a week at room temperature.

9. Sterilize the electrospun fibers by UV irradiation (30 min) followed by incubation in 70% ethanol for 15 min.
10. Wash scaffolds with sterilized deionized water three times to rinse away the ethanol.

3.2 Creating Adhesive Surfaces with pDOPA

1. To create the pDOPA adhesive surfaces for Cas9:sgRNA protein-RNA lipofectamine complexes to adhere onto, coat the electrospun scaffolds with mussel-inspired pDOPA.
2. Dissolve L-DOPA in DOPA polymerization buffer (Subheading 2.1), at a concentration of 0.5 mg/mL.
3. Wait 5 min for L-DOPA to be fully dissolved.
4. Incubate each electrospun scaffold in 1 mL of the DOPA solution on a shaker (80 rpm) for 2 h at room temperature.
5. After pDOPA coating, wash the electrospun fibers with sterilized deionized water to remove unbound L-DOPA monomers.
6. Coat laminin onto the scaffolds at 10 μg per scaffold (1 cm^2) for 2 h at 37 °C. Dilute laminin in sterile 1× PBS (*see* **Note 10**).

3.3 Formation of Cas9: sgRNA-Lipofectamine Complexes on the Scaffold

1. For Cas9:sgRNA complex formation, mix sgRNA and Cas9 protein in Opti-MEM at a molar ratio of 1:1 according to the manufacturer's protocol. Here, lipofectamine CRISPRMAX is used.
2. For a 1 cm^2 scaffold, dilute 7.5 pmol of sgRNA and Cas9 protein each in 25 μL of Opti-MEM.
3. Thoroughly mix the diluted sgRNA and Cas9 with 2.5 μL of Cas9 Plus reagent (provided with Lipofectamine CRISPRMAX).
4. Concurrently, dilute 1.5 μL of Lipofectamine CRISPRMAX into another microtube with 25 μL of Opti-MEM and incubate for 1 min at room temperature.
5. Add the mixture with sgRNA, Cas9 protein, and Cas9 Plus reagent (Cas9-sgRNA complexes) to the diluted Lipofectamine CRISPRMAX and incubate for 15 min at room temperature to form Cas9:sgRNA-lipofectamine complexes.
6. Transfer the mixture onto the scaffold (pDOPA-coated fibers). Incubate for 1 h at 37 °C.
7. Remove the supernatant and wash once with OptiMEM.
8. Seed cells at an appropriate density. For U2OS.EGFP, seed at 10,000 cells per cm^2 in 100 μL. Allow the cells to attach at 37 °C before submerging the whole scaffold with an additional 900 μL of culture medium.

3.4 Evaluation of EGFP Expression

1. After 3 days of culture, fix the cells on scaffolds with 4% PFA for 10 min at room temperature.
2. As cells tend to align and elongate along the aligned electrospun fibers, it may be difficult to detect the outline of cells. Therefore, additional staining for cellular cytoskeleton may be needed to count the number of cells for quantifying the changes in EGFP expression (*see* **Note 11**).
3. Add 0.1% Triton-X diluted in PBS to the scaffolds and incubate for 15 min.
4. Remove the supernatant and stain the cells with Rhodamine Phalloidin (1:500 in PBS) and DAPI (1:1000 in PBS) for 1 h at room temperature.
5. Wash the scaffolds three times with PBS.
6. Take images using an epifluorescence or confocal microscope.
7. Successful gene editing will result in knockout of EGFP. The knockout efficiency of EGFP can be quantified as the percentage of EGFP-positive cells per DAPI. Additional evaluation is needed to validate the genome editing efficiency (*see* **Note 12**).

3.5 Evaluation of Genomic Cleavage

1. After 3 days of culture, wash the scaffolds once with PBS.
2. Add 200 μL of Proteinase K solution to each scaffold and incubate the scaffolds in Proteinase K solution for overnight at 37 °C to extract genomic DNA.
3. Transfer the solution from scaffolds to the microtubes (one scaffold per tube). Flush the fibers with the solution a few times to ensure genomic DNA is washed out from the scaffolds (*see* **Note 13**).
4. Purify the genomic DNA using a DNA purification kit or established DNA purification protocol (*see* **Note 4**).
5. After the pure genomic DNA is obtained, run a genomic cleavage assay using a genomic cleavage detection kit or established protocol. During the PCR step to amplify the DNA, use the Forward and Reverse primers (Subheading 2.4) that are flanking the EGFP cleavage site (*see* **Notes 5** and **6**).
6. Successful genomic cleavage will result in two additional bands with DNA fragments of smaller sizes (~100 and 300 bp) in addition to the PCR product of EGFP (~400 bp). Quantify the cleavage efficiency based on the intensity of the bands on the gel using the ImageJ "Gel Analyzer" plugin.

4 Notes

1. Successful genome editing depends heavily on a good sgRNA design. The sequence here has been validated in multiple publications that target EGFP. For other genes, sgRNA has to be

designed. This can be done through various online sgRNA designing platforms. Once the sgRNA is designed or the sequence is known, it can be either synthesized in the lab using synthesis kits or ordered directly from companies that can synthesize the required sgRNA.

2. There are a variety of Cas9 proteins commercially available. We have only tested EnGen Cas9 NLS protein from New England Biolabs, which works well in this approach with U2OS.EGFP cells.
3. There is a range of commercially available delivery vehicles that are primarily lipid and cationic polymer-based. We have tested a few and found that lipofectamine CRISPRMAX works with high efficiency in our approach with USOS.EGFP cells. Note that this also depends on the types of cells being transfected as this may work differently for primary cells or difficult-to-transfect cells.
4. The topic of genomic DNA purification is beyond the scope of this chapter. Therefore, we only cover the methods until extracting genomic DNA from the scaffold efficiently. Established protocol or commercial DNA purification kits can be used to purify genomic DNA. We have tried Zymo DNA Clean & Concentrator and it worked well.
5. The topic of genomic cleavage assay is beyond the scope of this chapter. Established protocol or commercial genomic cleavage detection kits can be used to detect genomic cleavage. We have tried GeneArt Genomic Cleavage Detection Kit (Thermofisher) and it worked well. The primers here were designed for EGFP and sgRNA presented in this protocol. For other genes of interest, the primers shall be designed to flank the cleavage site that the Cas9:sgRNA complexes bind to.
6. The topic of DNA gel electrophoresis is beyond the scope of this chapter. We have used 2% agarose gel with TAE buffer. Alternatively, TBE buffer can be used. Other established protocols or commercial systems such as E-Gel can also be used to separate and detect the DNA fragments.
7. Ethidium bromide was used to visualize the DNA bands under UV light. Ethidium bromide was added to the 2% agarose gel to a final concentration of 0.5 μg/mL. Caution should be exercised as ethidium bromide is a known mutagen. Alternatively, other dyes such as SYBR Green can be used.
8. Here, we describe a method to fabricate electrospun fibers scaffold on the coverslips to simplify observation and imaging. A similar protocol may be applied to the other scaffold designs but optimization will be needed.

9. Aluminum foil may be used to make this mold with a dimension of 20 mm by 20 mm by 10 mm.
10. Laminin coating may be optional depending on the cell type seeded. This needs to be optimized. In our preliminary trial, we found that laminin enhanced cell proliferation of U2OS.EGFP cells on the scaffolds.
11. Confocal microscopy helps in clearly identifying GFP^+ and GFP^- cells with just GFP and DAPI signals on the fiber scaffolds. In this case, cytoskeleton staining may not be needed. Alternatively, other staining or methods that can visualize cellular outline may be used. However, we have tried using WGA to label cellular membranes, but it did not provide clear cellular outlines on the fiber scaffolds. Note that the fluorophore chosen shouldn't be in a similar range of wavelength that the GFP has.
12. For gene editing evaluation, genomic disruption assay using U2OS.EGFP is a fast and efficient method. However, for proper evaluation, genomic cleavage detection (GCD) assay is also needed to validate the results. Sequencing can also be performed to check for the indels. If a different scaffold design or when different reagents are used, loading efficiency and release profile may be checked through fluorescence tagged Cas9 or sgRNA. Alternatively, RNA assay (Ribogreen Assay) can be performed as a proxy for quantifying the sgRNA amount.
13. The optimal number of scaffolds per microtube depends on the required number of cells stated on the kit. Here, we used one scaffold (10, 000 U2OS.EGFP cells were seeded per scaffold and cultured for 3 days) per sample. If the protocol is modified or a different cell type is used, the number of scaffolds may need to be optimized. If more DNA is desired, pool solution from more scaffolds as one sample.

Acknowledgments

Partial funding support from the Singapore National Research Foundation under its National Medical Research Council-Cooperative Basic Research Grant (NMRC-CBRG) grant (NMRC/CBRG/0096/2015) and administered by the Singapore Ministry of Health's National Medical Research Council; Ministry of Education Tier 1 grant (RG38/19); and A*Star BMRC Singapore-China 12th Joint Research Programme Grant (Project No: 1610500024) are acknowledged. Jiah Shin Chin would like to thank the NTU Interdisciplinary Graduate Research Officer's scheme for supporting her through this work. We would also like to thank New England Biolabs for providing the Cas9 proteins.

References

1. Hsu PD, Lander ES, Zhang F (2014) Development and applications of CRISPR-Cas9 for genome engineering. Cell 157(6):1262–1278. https://doi.org/10.1016/j.cell.2014.05.010
2. Cox DB, Platt RJ, Zhang F (2015) Therapeutic genome editing: prospects and challenges. Nat Med 21(2):121–131. https://doi.org/10.1038/nm.3793
3. Perez-Pinera P, Kocak DD, Vockley CM, Adler AF, Kabadi AM, Polstein LR, Thakore PI, Glass KA, Ousterout DG, Leong KW, Guilak F, Crawford GE, Reddy TE, Gersbach CA (2013) RNA-guided gene activation by CRISPR-Cas9-based transcription factors. Nat Methods 10(10):973–976. https://doi.org/10.1038/nmeth.2600
4. Damian M, Porteus MH (2013) A crisper look at genome editing: RNA-guided genome modification. Mol Ther 21(4):720–722. https://doi.org/10.1038/mt.2013.46
5. Chakraborty S, Ji H, Kabadi AM, Gersbach CA, Christoforou N, Leong KW (2014) A CRISPR/Cas9-based system for reprogramming cell lineage specification. Stem Cell Rep 3(6):940–947. https://doi.org/10.1016/j.stemcr.2014.09.013
6. Li L, Hu S, Chen X (2018) Non-viral delivery systems for CRISPR/Cas9-based genome editing: challenges and opportunities. Biomaterials 171:207–218. https://doi.org/10.1016/j.biomaterials.2018.04.031
7. Wang HX, Li M, Lee CM, Chakraborty S, Kim HW, Bao G, Leong KW (2017) CRISPR/Cas9-based genome editing for disease modeling and therapy: challenges and opportunities for nonviral delivery. Chem Rev 117(15):9874–9906. https://doi.org/10.1021/acs.chemrev.6b00799
8. De Laporte L, Shea LD (2007) Matrices and scaffolds for DNA delivery in tissue engineering. Adv Drug Deliv Rev 59(4–5):292–307. https://doi.org/10.1016/j.addr.2007.03.017
9. Whitaker MJ, Quirk RA, Howdle SM, Shakesheff KM (2001) Growth factor release from tissue engineering scaffolds. J Pharm Pharmacol 53(11):1427–1437. https://doi.org/10.1211/0022357011777963
10. Low WC, Rujitanaroj PO, Lee DK, Kuang J, Messersmith PB, Chan JK, Chew SY (2015) Mussel-inspired modification of nanofibers for REST siRNA delivery: understanding the effects of gene-silencing and substrate topography on human mesenchymal stem cell neuronal commitment. Macromol Biosci 15(10):1457–1468. https://doi.org/10.1002/mabi.201500101
11. Low WC, Rujitanaroj PO, Lee DK, Messersmith PB, Stanton LW, Goh E, Chew SY (2013) Nanofibrous scaffold-mediated REST knockdown to enhance neuronal differentiation of stem cells. Biomaterials 34(14):3581–3590. https://doi.org/10.1016/j.biomaterials.2013.01.093
12. Chooi WH, Ong W, Murray A, Lin J, Nizetic D, Chew SY (2018) Scaffold mediated gene knockdown for neuronal differentiation of human neural progenitor cells. Biomater Sci 6(11):3019–3029. https://doi.org/10.1039/c8bm01034j
13. Zhang N, Milbreta U, Chin JS, Pinese C, Lin J, Shirahama H, Jiang W, Liu H, Mi R, Hoke A, Wu W, Chew SY (2019) Biomimicking fiber scaffold as an effective in vitro and in vivo microRNA screening platform for directing tissue regeneration. Adv Sci (Weinh) 6(9):1800808. https://doi.org/10.1002/advs.201800808
14. Ong W, Lin J, Bechler ME, Wang K, Wang M, Ffrench-Constant C, Chew SY (2018) Microfiber drug/gene delivery platform for study of myelination. Acta Biomater 75:152–160. https://doi.org/10.1016/j.actbio.2018.06.011
15. Diao HJ, Low WC, Lu QR, Chew SY (2015) Topographical effects on fiber-mediated microRNA delivery to control oligodendroglial precursor cells development. Biomaterials 70:105–114. https://doi.org/10.1016/j.biomaterials.2015.08.029
16. Diao HJ, Low WC, Milbreta U, Lu QR, Chew SY (2015) Nanofiber-mediated microRNA delivery to enhance differentiation and maturation of oligodendroglial precursor cells. J Control Release 208:85–92. https://doi.org/10.1016/j.jconrel.2015.03.005
17. Chin JS, Chooi WH, Wang H, Ong W, Leong KW, Chew SY (2019) Scaffold-mediated non-viral delivery platform for CRISPR/Cas9-based genome editing. Acta Biomater 90:60–70. https://doi.org/10.1016/j.actbio.2019.04.020

Chapter 14

Correction of Hemoglobin E/Beta-Thalassemia Patient-Derived iPSCs Using CRISPR/Cas9

Methichit Wattanapanitch

Abstract

HbE/β-thalassemia is one of the most common thalassemic syndromes in Southeast Asia and Thailand. Patients have mutations in β hemoglobin (*HBB*) gene resulting in decreased and/or abnormal production of β hemoglobin. Here, we describe a protocol for CRISPR/Cas9-mediated gene correction of the mutated hemoglobin E from one allele of the *HBB* gene by homology-directed repair (HDR) in HbE/β-thalassemia patient-derived induced pluripotent stem cells (iPSCs) using a CRISPR/Cas9 plasmid-based transfection method and a single-stranded DNA oligonucleotide (ssODN) repair template harboring the correct nucleotides. Our strategy allows the seamless *HbE* gene correction with the editing efficiency (HDR) up to 3%, as confirmed by Sanger sequencing. This protocol provides a simple one-step genetic correction of *HbE* mutation in the patient-derived iPSCs. Further differentiation of the corrected iPSCs into hematopoietic stem/progenitor cells will provide an alternative renewable source of cells for the application in autologous transplantation in the future.

Key words Beta-thalassemia, Hemoglobinopathy, Genetic correction, Induced pluripotent stem cells, CRISPR/Cas9, Homology-directed repair (HDR), Clonal isolation

1 Introduction

Thalassemia is an inherited autosomal recessive disease characterized by decreased production or structural alteration of the α or β hemoglobin chain leading to improper oxygen transport and hemolytic anemia [1, 2]. The complexity of thalassemia syndromes is a result of interactions between different combinations of globin gene defects. Several different combinations of abnormal genes can result in over 60 thalassemic syndromes.

One of the most common syndromes in Thailand and Southeast Asia is HbE/β-thalassemia, which has heterogeneous clinical manifestations from very mild and asymptomatic anemia to very severe disorder requiring transfusions. Patients with HbE/β-thalassemia can have various mutations in one allele (β^0 or β^+) producing no or decreased β-globin chain, and a missense mutation at

Kumaran Narayanan (ed.), *Bio-Carrier Vectors: Methods and Protocols*, Methods in Molecular Biology, vol. 2211,
https://doi.org/10.1007/978-1-0716-0943-9_14,

codon 26 (GAG → AAG, glutamate to lysine) in another allele (β^E) resulting in abnormal hemoglobin E production [3]. Allogeneic hematopoietic stem cell transplantation is the only curative therapy. However, the treatment is limited by the availability of HLA-matched donors.

The advent of iPSC technology provides a patient-derived renewable source of cells amenable to genetic manipulation. Generation of patient-derived iPSCs followed by genetic correction of mutations and differentiation into specific cell types offer promise for autologous transplantation. Previously, gene therapy using a lentiviral vector carrying β-globin (*HBB*) gene has been reported in β-thalassemia patient-derived iPSCs [4, 5]. However, this approach can lead to insertional mutagenesis and therefore requires the screening of iPSC clones that harbor the *HBB* transgene at genomic safe harbor locus.

Recently, the RNA-guided CRISPR/Cas9 system has become a versatile tool for precise genome editing and been used to efficiently correct the mutations in homozygous β-thalassemia patient-derived iPSCs through homology-directed repair (HDR) pathway [6–9]. In this chapter, we describe the detailed protocol for the seamless correction of *HbE* mutation in HbE/β-thalassemia patient-derived iPSCs (Eβ-iPSCs) using the CRISPR/Cas9 plasmid and the ssODN repair template according to our previous work [10]. The gene-targeting strategy involves designing of the gRNA targeting the point mutation (**A**AG) and the ssODN template containing the left homology arm, the correct nucleotide (**G**AG) and the right homology arm, and construction and cloning of the gRNA into the Cas9 expression plasmid (PX459) to obtain the CRISPR/Cas9 plasmid (sgRNA/PX459). To target the *HbE* mutation, we perform nucleofection to deliver both the sgRNA/PX459 and the ssODN template into the Eβ-iPSCs. The transfected pools are clonally isolated and expanded for verification of the corrected clones by multiplex PCR analysis and Sanger sequencing (Fig. 1).

After genetic correction of the *HbE* mutation in one allele, the corrected iPSCs become heterozygote. Upon hematopoietic and erythroid differentiation, the corrected iPSCs can restore HBB gene and protein expression [10]. This genetic correction strategy is simpler and more practical than correcting the *HBB* mutations in the other allele, which can be heterogeneous with over 200 possible mutations at the *HBB* gene being identified.

Therefore, the strategy can be employed as a universal approach for *HbE* correction of other types of HbE/β^0 or HbE/β^+-thalassemia patient-derived iPSCs. This protocol is also applicable to genome editing of iPSCs for creating small edits such as correction of point mutations in iPSCs derived from patients with genetic disorders or introduction of point mutations into the wild-type iPSCs to create the isogenic diseased iPSC lines for functional genomics study.

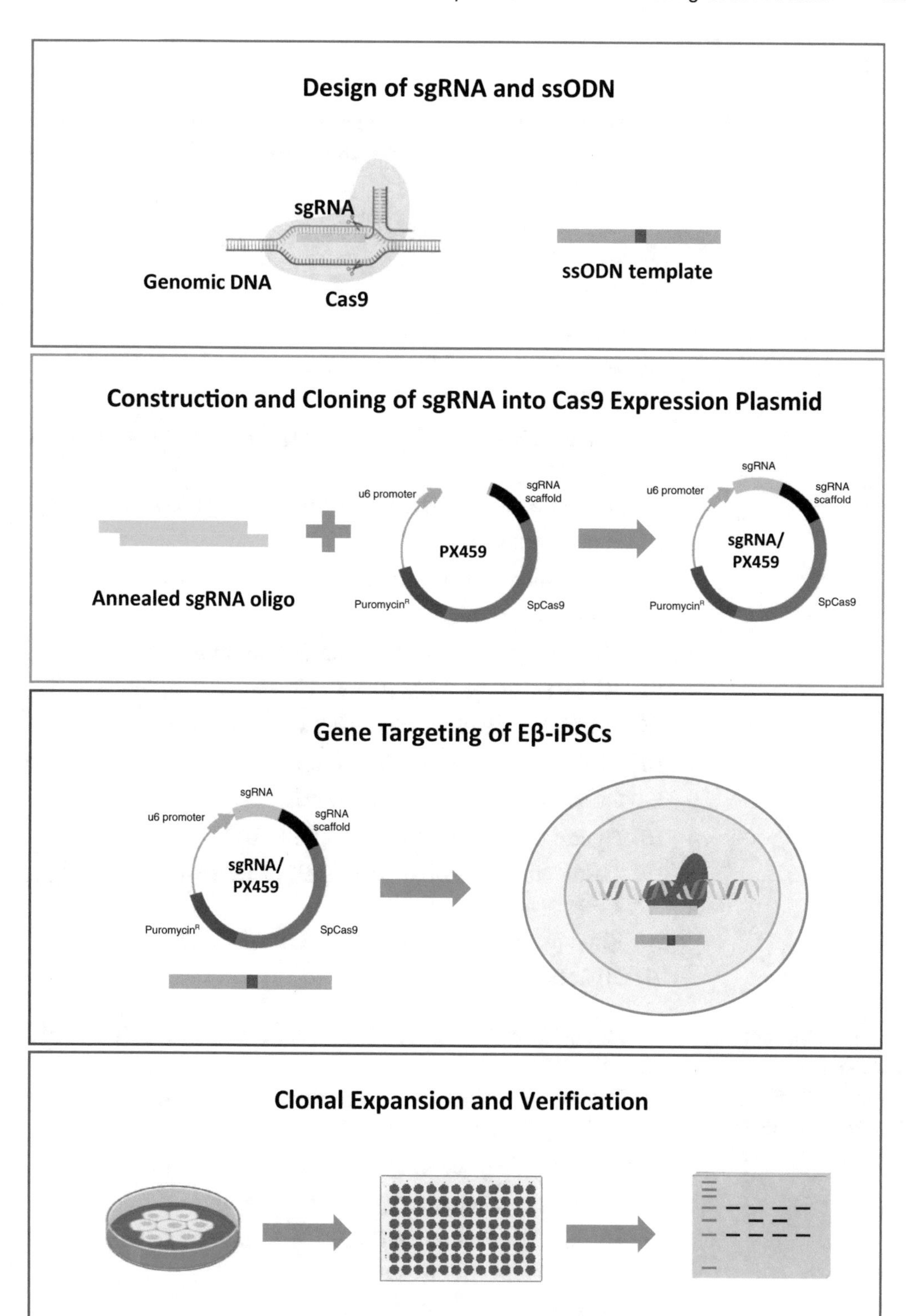

Fig. 1 Schematic diagram of genome editing in Eβ-iPSCs

2 Materials

2.1 Culture of iPSCs

1. We used Eβ-iPSCs for the correction of HbE mutation. The cell line was derived from skin fibroblasts of β^E/β^0-thalassemia patients [10].
2. Accutase (Merck Millipore).
3. Cell scraper.
4. Cryogenic box.
5. Cryogenic vials (2 mL).
6. Dispase (StemCell Technologies).
7. DMSO.
8. Dulbecco's Modified Eagle Medium/F12 (DMEM/F12) (Gibco).
9. Dulbecco's PBS without Calcium and Magnesium (Gibco).
10. Hemocytometer.
11. Humidified CO_2 incubator.
12. Liquid nitrogen tank.
13. Matrigel® hESC-qualified Matrix (Corning®).
14. Microcentrifuge tubes (0.5 and 1.5 mL).
15. mTeSR™1 medium (StemCell Technologies).
16. 15-mL and 50-mL conical tubes.
17. Sterile seropipettes (5 and 10 mL).
18. Pipette tips (2, 20, 200, and 1000 μL).
19. Tissue culture-treated multi-well plates (6-well, 12-well, and 24-well plates).
20. Rho kinase inhibitor Y-27632 (Calbiochem).
21. 70% ethanol.

2.2 Construction and Cloning of Single-Guide (sgRNA) into Cas9 Expression Plasmid

1. 1% Agarose: 1 g of agarose powder dissolved in 100 mL of 1× TBE buffer.
2. 100 μg/mL Ampicillin: working concentration is 100 mg/mL in ddH_2O. Store at −20 °C.
3. 10 mM ATP (New England BioLabs).
4. Chemically competent DH5α *E. coli* cells.
5. 10 mM Dithiothreitol (DTT): 1.5 mg of DTT powder dissolved in 1 mL of ddH_2O.
6. Gel chamber and power source.
7. Fast digest *Bbs*I restriction enzyme (Thermo Fisher Scientific).

8. Fast digest *Eco*RI restriction enzyme (Thermo Fisher Scientific).
9. Gel documentation machine.
10. GenepHlow™ gel extraction and PCR Cleanup kit (Geneaid).
11. 100% glycerol.
12. Ice.
13. Incubating shaker.
14. LB-ampicillin agar: 35 g/L of LB agar, Lennox and 100 μg/mL ampicillin.
15. LB-ampicillin broth: 25 g/L of Difco™ LB broth and 100 μg/mL ampicillin.
16. Molecular grade water (ddH_2O).
17. NanoDrop 2000 (Thermo Fisher Scientific).
18. Novel juice (DNA staining reagent) (GeneDireX).
19. PCR machine T100™ thermal cycler (Bio-Rad).
20. PCR tubes (0.2 mL).
21. pSpCas9(BB)-2A-Puro (PX459) (Addgene) contains the U6 promoter, sgRNA scaffold, *Sp*Cas9 and puromycin-resistant gene.
22. Sense and antisense strands of sgRNA oligonucleotides with *Bbs*I overhangs (Table 1).
23. Cell spreader.
24. Standard ladder Lambda DNA/HindIII Marker (Thermo Fisher Scientific).
25. T4 ligation buffer (Thermo Fisher Scientific).
26. T4 Polynucleotide Kinase (PNK) (Thermo Fisher Scientific) .
27. T7 ligase (New England BioLabs).
28. 10× Tango buffer (Thermo Fisher Scientific).
29. 1× TBE buffer: 0.089 M Tris, 0.089 M Borate, 0.002 M EDTA.

Table 1
Oligonucleotide sequences for sgRNA construction

Strand	Oligo sequence (5′ to 3′)	PAM
Sense	**CACC***G*TGGATGAAGTTGGTGGTA	AGG
Antisense	**AAAC**TACCACCAACTTCATCCA*C*	

Table 2
Primer sequence for U6 promoter

Primer name	Oligo sequence (5′ → 3′)	Length (base)	Ref.
U6-forward primer	GAGGGCCTATTTCCCATGATTCC	23	[11]

30. U6 promoter forward primer (Table 2).
31. Temperature-controlled water bath.
32. Presto™ Mini Plasmid Kit, Geneaid™ Midi Plasmid Kit (Geneaid).

2.3 Gene Targeting of Eβ-iPSCs

1. Amaxa 4D-nucleofector System (Lonza).
2. P3 Primary Cell 4D-Nucleofector™ X Kit S (Lonza). This kit contains Nucleofector™ Solution, Supplement, and 16-well Nucleocuvette™ Strips (20 μL).
3. sgRNA/PX459 plasmid, used for gene targeting of the E-β-iPSCs: construction described in Subheading 3.3.
4. Single-stranded DNA oligonucleotide (ssODN) repair template. The ssODN repair template for *HbE* correction contains the homology arms of 90 nucleotides on either side of the point mutation (a total of 181 nucleotides).
   (a) Order the ssODN repair template according to the sequence below (*see* **Note 1**). The nucleotide "G" (bold letter) represents the right nucleotide. CAAACAGACAC CATGGTGCATCTGACTCCTG AGGAGAAGTCTGCCGTTACTGCCC TGTGGGGCAAGGTGAACGTGGAT GAAGTTGGTGGT**G**AGGCCCTGGGCAGGTTGG TATCAAGGTTACAAGACA GGTTTAAGGAGACCAATAGAAACTG GGCATGTGGAGACAGAGAAGACTCTTGGG.
   (b) Resuspend the ssODN to 200 μM in sterile ddH_2O and store at −20 °C.

2.4 Clonal Isolation of Genetic Corrected iPSCs

1. 40-μm cell strainer.
2. 96-well plates.
3. SMC4 small molecule cocktail (Corning).

2.5 Verification of Clones by Multiplex PCR Analysis

1. Heat block.
2. HotStarTaq polymerase, PCR buffer, $MgCl_2$, dNTPs (Qiagen).
3. Primers for multiplex PCR analysis for HbE mutation (Table 3).

Table 3
Primer sequences for multiplex PCR analysis of hemoglobin E

Primer name	Oligo sequence (5′→3′)
HbE-Fc	TCCAACTCCTAAGCCAGTGC
HbE-Rn	CCTGCCCAGGGCCTC
HbE-Fm	CGTGGATGAAGTTGGTGGTA
HbE-Rc	CGATCCTGAGACTTCCACACTG

Table 4
Primer sequences for off-target analysis

OT	Forward primer (5′→3′)	Reverse primer (5′→3′)	Product size (bp)
1	AACCAACCTGCTCACTGGAG	AGCCTTCACCTTAGGGTTGC	309
2	TGGAAGCAGGTGGACAGTTC	TGTGTAGGTTACCCAAGGCAC	369
3	TGCAGGTGTGTAGGTTGAGTC	CCAAACAACCCCTACCACCA	352
4	AACCGCATGGAGTCGTTCTT	ATGGCCCCGAACTAACAGTG	307
5	GAGTTGACCCGAAGAAGCTG	GAGTTGACCCGAAGAAGCTG	412

4. QuickExtract™ DNA Extraction Solution (Epicentre).
5. Vortex.

2.6 Off-Target Analysis

1. Primers for top five off-targets (Table 4).
2. Q5 High-Fidelity DNA polymerase, Q5 PCR buffer with $MgCl_2$ and dNTPs (New England Biolabs).
3. QuickExtract™ DNA Extraction Solution (Epicentre).

3 Methods

The following section provides detailed protocols that we use for the correction of *HbE* mutation in the Eβ-iPSCs. The protocols include the culture of iPSCs, design and construction of sgRNA, cloning of sgRNA into Cas9 expression plasmid, design and preparation of the ssODN repair template, gene targeting of *HbE* mutation, clonal selection, and verification.

3.1 Culture of iPSCs

iPSCs are maintained under a feeder-free condition in mTeSR™1 medium on Matrigel-coated plates.

3.1.1 Thawing Cryopreserved iPSCs

1. Warm mTeSR™1 medium to room temperature (*see* **Note 2**).

2. Pre-coat the 6-well culture plate with 1 mL of the diluted Matrigel (*see* **Note 3**) and leave the plate at room temperature for an hour or at 37 °C for at least 30 min.
3. Quickly thaw cryopreserved iPSCs in a 37 °C water bath until only a small ice crystal remains.
4. Wipe the outside of the cryovial with 70% ethanol.
5. Gently pipette the cell suspension and transfer to a 15-mL conical tube.
6. Add 5 mL mTeSR™1 medium dropwise to the cell suspension while gently rocking the tube.
7. Centrifuge the cells at 500 × *g* for 5 min at room temperature.
8. Aspirate the diluted Matrigel from the well and add 1 mL mTeSR™1 medium supplemented with 10 μM Y-27632.
9. After centrifugation, aspirate the supernatant.
10. Gently resuspend the cells in 1 mL mTeSR™1 medium supplemented with 10 μM Y-27632. Take care to maintain the cells as clumps for better attachment.
11. Transfer the cell suspension onto the Matrigel-coated well (step 8) and incubate at 37 °C, 5% CO_2.
12. After 24 h, replace the medium with mTeSR™1 medium without Y-27632 and change the medium daily.

3.1.2 Passaging iPSCs

Passage when the cells reach 70% confluence, usually every 4–5 days (*see* **Note 4**).

1. Pre-coat the 6-well culture plate with 1 mL of the diluted Matrigel (*see* **Note 3**) for an hour at room temperature. Aspirate the diluted Matrigel from the well and add 2 mL of mTeSR™1 medium.
2. Discard the medium from the cultured well and wash the cells with DMEM/F12.
3. Add 1 mL of 1 mg/mL Dispase and incubate for 7 min at 37 °C or until the edge of the colony is rounded up.
4. Aspirate Dispase, wash the well twice with DMEM/F12.
5. Add 2 mL of mTeSR™1 medium, gently scrape off the colonies using cell scraper.
6. Use a 5-mL seropipette, gently break up large clumps into small clumps.
7. Transfer 200 μL of the cell clumps onto the Matrigel-coated well containing 2 mL of mTeSR™1 medium (**step 1**) and incubate at 37 °C, 5% CO_2.
8. Change the medium daily.

3.1.3 Cryopreserving iPSCs

This protocol is suitable for a confluent well of a 6-well plate.

1. Prepare 5 mL of a freezing solution (10% DMSO in mTeSR™1 medium) and chill at 4 °C.
2. Harvest iPSCs according to the passaging protocol described in Subheading 3.1.2, **steps 2–6**. Minimize the break-up of cell aggregates.
3. Centrifuge the cell suspension at 500 × *g* for 5 min at room temperature.
4. Aspirate the medium.
5. Add 4 mL of chilled freezing solution dropwise to the cells while gently rocking the tube back and forth.
6. Quickly dispense 1 mL aliquots of the suspension into four cryogenic vials.
7. Transfer the cryovials into the cryopreservation box and store at −80 °C freezer for overnight.
8. Next day, transfer the vials to a liquid nitrogen tank for storage.

3.2 Design and Construction of sgRNA

The protocol for the construction of sgRNA is adapted from Ran, et al. [11].

1. Design sgRNAs targeting the *HbE* mutation using the CRISPR Design Tool (http://crispr.mit.edu/). Input the region of around 200 bp spanning the *HbE* point mutation. Select the gRNAs with the highest specificity and the lowest off-targets, and the cleavage site as close to the *HbE* mutation as possible (*see* **Note 5**).
2. Order the 20-bp oligonucleotide sequences as shown in Table 1 (*see* **Note 6**).
3. Dilute the sgRNA oligonucleotides by resuspending the sense and antisense strands in ddH_2O to a final concentration of 100 μM.
4. Set up the reaction for phosphorylating and annealing of the sgRNA oligonucleotides (Table 5).

Table 5
Components for phosphorylation and annealing of the sgRNA oligos

Components	Amount (μL)
sgRNA sense (100 μM)	1
sgRNA antisense (100 μM)	1
T4 ligation buffer (10×)	1
T4 PNK	1
ddH_2O	6
Total	**10**

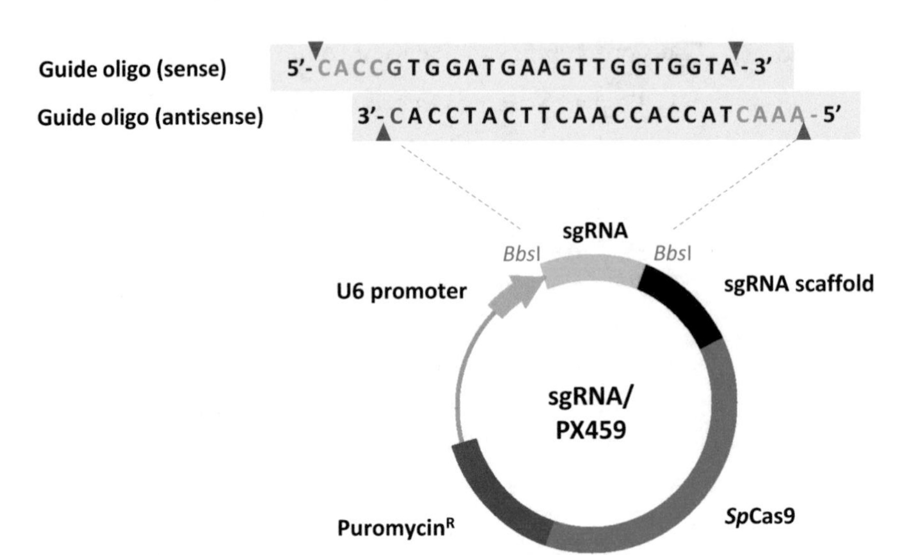

Fig. 2 Schematic diagram shows cloning of the annealed sgRNA oligo into the PX459 plasmid. The oligos contain overhangs for ligation (blue) into the pair of *Bbs*I sites of the PX459 plasmid. A G-C base pair (red) is added at the 5′ end of the sgRNA sequence

5. Place the reaction tube in a thermocycler by using the following parameters: 37 °C for 30 min; 95 °C for 5 min; ramp down to 25 °C at 5 °C/min.
6. Dilute the phosphorylated and annealed oligonucleotide 1:200 by adding 1 μL of the oligonucleotide (**step 5**) to 199 μL of ddH_2O.

3.3 Cloning of the Annealed sgRNA Oligo into PX459 (sgRNA/PX459)

Clone the annealed sgRNA oligo into the *Bbs*I restriction site of the PX459 plasmid to obtain the sgRNA/PX459 plasmid (Fig. 2).

3.3.1 Ligation of sgRNA Oligonucleotides

1. Set up a ligation reaction for the sgRNA (Table 6). For negative control, prepare a no-insertion, PX459-only for ligation.
2. Incubate the ligation reaction in a thermal cycler using 6 cycles of this condition: 37 °C for 5 min and 21 °C for 5 min.

3.3.2 Transformation of Cloned sgRNA/PX459 Plasmid

1. Thaw a chemically competent *E. coli* strain, DH5α cells, slowly on ice.
2. Transform 2 μL of the ice-cold ligation mixture (Subheading 3.3.1) to 50 μL of the ice-cold competent DH5α cells in a 1.5-mL microfuge tube and incubate on ice for 10 min.
3. Heat shock the cells in a water bath at 42 °C for 45 s.
4. Immediately return the microfuge tube to ice for 5 min and add 900 μL of LB medium without ampicillin.

Table 6
Components for ligation

Components	Amount (μL) for 1 reaction
PX459, 100 ng	X
Diluted oligo duplex (Subheading 3.2, **step 6**)	2
Tango buffer, 10×	2
DTT, 10 mM	1
ATP, 10 mM	1
FastDigest *Bbs*I	1
T7 ligase	0.5
ddH_2O	Up to 20
Total	**20**

5. Incubate the microfuge tube in the shaker at 37 °C for 1 h at 250 rpm.
6. Centrifuge at 2000 × *g* for 5 min.
7. Aspirate 500 μL of the LB medium, resuspend the cell pellet, and spread 150 μL of cell suspension onto an LB agar plate containing 100 μg/mL ampicillin.
8. Incubate at 37 °C overnight (*see* **Note 7**).
9. Pick up 3–5 single colonies, inoculate 6 mL of LB medium supplemented with 100 μg/mL ampicillin, incubate in the shaker at 37 °C at 250 rpm overnight.
10. Aliquot 5 mL of culture for plasmid extraction using Presto™ Mini Plasmid Kit, following the manufacturer's instruction and store 1 mL of culture at 4 °C for short-term storage. Measure the concentration and quality of the sgRNA/PX459 plasmid using NanoDrop 2000. Store the plasmid at −20 °C.

3.3.3 Sequence Validation of CRISPR Plasmid

At Least 100 Ng of the Recombinant plasmid Is Required for Sequencing.

1. Verify the sequence of the isolated sgRNA/PX459 plasmid by sequencing from the U6 promotor using the U6-Forward primer (Table 2).
2. Select the bacterial culture with the correct DNA sequence for plasmid preparation using Geneaid™ Midi Plasmid Kit for the nucleofection step. Add 1 mL of culture that has been stored at 4 °C (from Subheading 3.3.2, **step 10**) into fresh 100 mL of LB medium containing 100 μg/mL ampicillin and incubate in the shaker at 37 °C at 250 rpm overnight.

3. Perform plasmid extraction using 50 mL of culture with Geneaid™ Midi Plasmid Kit according to the manufacturer's instruction (*see* **Note 8**).
4. Measure the plasmid concentration using NanoDrop 2000 and store the plasmid at −20 °C. This preparation will be used for the nucleofection step in Subheading 3.4.
5. Store the culture in 50% glycerol at −80 °C.

3.4 Gene Targeting of Eβ-iPSCs

Genetic correction of *HbE* mutation in the Eβ-iPSCs is performed using the sgRNA/PX459 plasmid generated in Subheading 3.3.3 and the ssODN repair template (Subheading 2.3) on the Amaxa 4D-Nucleofector System. The protocol is adapted from previously described steps [11, 12].

1. Culture the Eβ-iPSCs in mTeSR™1 medium on a Matrigel-coated 6-well plate (*see* **Note 3**) for 3–4 days or until they reach 50–70% confluence.
2. On the day of transfection, pre-treat the Eβ-iPSCs with 10 μM Y-27632 for an hour.
3. Prepare a Matrigel-coated 24-well plate for seeding (*see* **Note 3**).
4. Remove the mTeSR™1 medium from the 6-well cultured plate (**step 1**) and wash the cells once with DPBS.
5. Add 1 mL/well Accutase and incubate at 37 °C for 7 min.
6. Add 1 mL/well mTeSR™1 medium supplemented with 10 μM Y-27632, count the cells (200,000 cells per reaction is required), centrifuge the cells in a 15-mL conical tube at 500 × *g* for 5 min at room temperature.
7. Prepare the P3 Primary Cell Nucleofector™ solution (Subheading 2.3, **step 2**) by mixing 3.6 μL of the supplement with 16.4 μL of the Nucleofector™ solution in a 1.5-mL microfuge tube.
8. Remove the medium (**step 6**) completely and resuspend 200,000 cells in 20 μL of the P3 Primary Cell Nucleofection™ solution (**step 7**).
9. Add 2 μg of the sgRNA/pX459 plasmid and 200 pmol ssODN repair template (total volume does not exceed 2 μL) into the cell suspension.
10. Transfer the mixture to one well of the 16-well Nucleocuvette™ Strip (*see* **Note 9**) and run the program CB-150 according to the manufacturer's instruction.
11. Immediately add 80 μL of mTeSR™1 medium supplemented with 10 μM Y-27632 into the well and transfer the cells into the 24-well Matrigel-coated plate (from **step 3**) containing 1 mL of pre-warmed mTeSR™1 medium supplemented with 10 μM Y-27632.

12. Incubate the transfected cells at 37 °C, 5% CO_2.
13. Refeed the cells daily with mTeSR™1 medium after 24 h post-nucleofection.

3.5 Clonal Isolation of Genetic Corrected iPSCs

Clonal isolation of the transfected iPSCs can be performed as early as 24 h post-transfection by FACS or serial dilutions. This section provides the protocol for clonal isolation using serial dilutions. We obtain better viability using serial dilutions as compared to FACS.

1. At day 3 post-transfection, pre-treat the transfected cells with SMC4 in mTeSR™1 medium for an hour (*see* **Note 10**).
2. Prepare Matrigel-coated 96-well plates for seeding (*see* **Note 3**).
3. Remove the mTeSR™1 medium from the 24-well cultured plate (**step 1**) and wash the cells once with DPBS.
4. Dissociate the cells using 0.25 mL of Accutase at 37 °C for 7 min.
5. Add 0.25 mL of mTeSR™1 medium supplemented with SMC4.
6. Centrifuge the cells at 500 × *g* for 5 min at room temperature. Filter the cells using 40-μm cell strainer.
7. Resuspend the cells in mTeSR™1 medium supplemented with SMC4. Dilute the cells and seed the cells at a density of 10 cells in 100 μL/well of 96-well plates (*see* **Note 11**).
8. Seed at least two 96-well plates and incubate at 37 °C, 5% CO_2.
9. The next day, examine the plate under a microscope. Identify the well that contains a single cell and mark off the wells that are empty or contain multiple cells.
10. Culture the cells in mTeSR™1 medium supplemented with SMC4 for 8 days.
11. Expand the cells for 2 weeks or until the size of the colony is large enough for picking.
12. Manually pick half of the single colony using a 200-μL pipette tip for genomic DNA isolation and genotyping (*see* Subheading 3.6).
13. To establish a back-up culture, pick up the remaining half of the colony and transfer to a well of a Matrigel-coated 12-well plate (*see* **Note 3**) containing the pre-warmed mTeSR™1 medium supplemented with 10 μM Y-27632.

3.6 Verification of Clones by Multiplex PCR Analysis of Hemoglobin E

The following protocol describes isolation of genomic DNA from the iPSC clones for genotyping after gene editing.

1. Transfer half of the colony from Subheading 3.5 (**step 12**) to a 0.5-mL tube containing 10 μL of QuickExtract™ solution.

2. Mix by vortexing for 15 s and incubate in a heat block at 65 °C for 6 min.
3. Mix by vortexing for 15 s and incubate in a heat block at 98 °C for 2 min.
4. PCR amplify the genomic region spanning the Cas9 target site (*HbE* point mutation) using the primers described in Table 3 (and Fig. 3) using the PCR reaction in Table 7.
5. Transfer the PCR tubes to a thermal cycler using conditions listed in Table 8.
6. After PCR, run the samples on 1% (w/v) agarose gel at 100 V for 25 min.
7. Select the iPSC colonies with HbE negative based on the PCR result (Fig. 4).

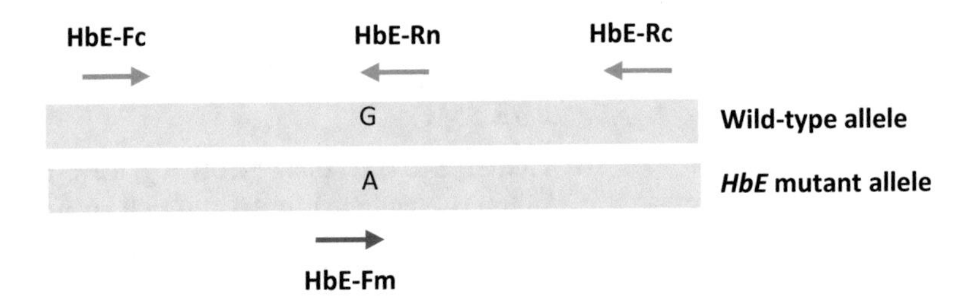

Fig. 3 Primer design for multiplex PCR analysis

Table 7
Reaction setup for multiplex PCR analysis of hemoglobin E

Component	Amount (μL)	Final concentration
HotStarTaq polymerase, 5 units/μL	0.3	1.5 U
PCR buffer, 10×	3	1×
$MgCl_2$, 25 mM	1.8	1.5 mM
dNTPs, 2 mM	1.5	0.1 mM
HbE-Fc primer, 10 μM	1.2	0.4 μM
HbE-Rc primer, 10 μM	1.2	0.4 μM
HbE-Rn primer, 10 μM	1.5	0.5 μM
HbE-Fm primer, 10 μM	1.5	0.5 μM
DNA sample	2	
Nuclease-free water	To 30	

Table 8
Thermal cycling conditions for PCR verification of clones

Step	Temperature and time
Initial denaturation	95 °C, 15 min
30 cycles	94 °C, 45 s 68 °C for 45 s 72 °C for 1 min
Final extension	72 °C for 7 min
Hold	4 °C

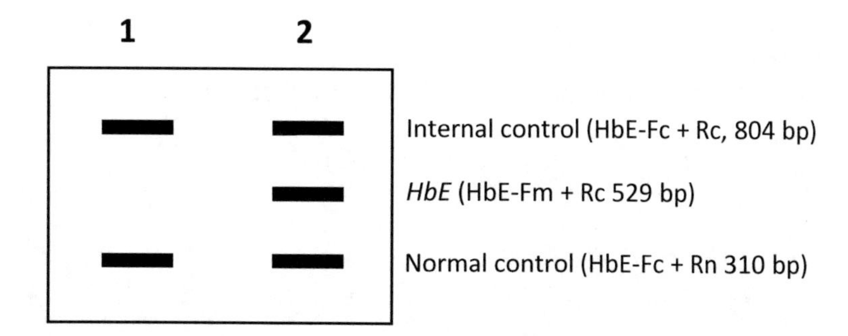

Fig. 4 Expected results for multiplex PCR analysis of hemoglobin E

8. Purify the PCR product using the GenepHlow™ gel/PCR kit according to the manufacturer's instruction and further confirm the result with Sanger sequencing.
9. Expand the HbE negative clones and cryopreserve (Subheading 3.1.3) for further analyses.

3.7 Off-Target Analysis

For off-target analysis, perform PCR amplification of five potential off-target sites on genomic DNA of the corrected iPSC clones.

1. Extract genomic DNA from the corrected iPSC clones (Subheading 3.6, **step 9**) using the QuickExtract™ solution.
2. Perform PCR for off-target analysis using the isolated genomic DNA template, the primers (Table 4), and the PCR reaction described in Table 9.
3. Transfer the PCR tubes to a thermal cycler using conditions listed in Table 10.
4. After PCR, run the samples on 1% (w/v) agarose gel at 100 V for 25 min.

Table 9
Reaction setup for off-target analysis

Component	Amount (μL)	Final concentration
Q5 high-Fidelity DNA polymerase, 100×	0.5	0.2 U
Q5 reaction buffer with $MgCl_2$, 5×	10	1×
dNTPs, 10 mM	1	0.2 mM
Each primer, 10 μM	2.5	0.5 μM
DNA sample	2	
Nuclease-free water	To 50	

Table 10
Thermal cycling conditions for PCR verification of clones

Step	Temperature and time
Initial denaturation	98 °C, 30 s
35 cycles	98 °C, 10 s 68 °C for 30 s 72 °C for 30 sec
Final extension	72 °C for 2 min
Hold	4 °C

5. Purify the PCR product by using the GenepHlow™ gel/PCR kit according to the manufacturer's instruction and further confirm the result with Sanger sequencing.
6. Analyze the sequencing data using the read alignment programs such as MUSCLE (https://www.ebi.ac.uk/Tools/msa/muscle/).

4 Notes

1. The length of homology arms of the ssODN repair template can range from 90–180 nucleotides. The ssODN repair template should be HPLC purified.
2. Thaw the mTeSR™1 5× supplement at room temperature or 4 °C for overnight. Do not thaw the supplement at 37 °C. Prepare complete mTeSR™1 medium by mixing the mTeSR™1 5× supplement and the mTeSR™1 basal medium. Store the complete mTeSR™1 medium at 2–8 °C for up to

2 weeks. Alternatively, aliquot complete mTeSR™1 medium and store at −20 °C for up to 6 months. Do not freeze-thaw the complete medium.

3. Always keep Matrigel on ice when thawing and handling to prevent gelation. The optimal working concentration of Matrigel is cell line dependent. We recommend diluting Matrigel in ice-cold DMEM/F12 medium at 1:30 to 1:40 dilutions. Matrigel-coated plates can be used immediately or sealed with Parafilm to prevent evaporation and stored at 2–8 °C for up to a week. Prior to use, pre-warm the culture plate to room temperature for at least 1 h.

4. Passaging is routinely performed every 4–5 days depending on the cell line, size of colonies, and plating density. Maintaining appropriate colony size and density is very important for keeping a good quality of iPSC culture. Cultures that are too dense can result in spontaneous differentiation where the colonies lose their border integrity and contain regions of irregular shape or other cell types. If that occurs, remove the differentiated regions using pipette tip under the inverted microscope.

5. We use the *Sp*Cas9 which cuts at 3 nucleotides upstream of the PAM (5′-NGG). Ideally, the cleavage site should be within 10 nucleotides from the mutation to obtain the best targeting efficiency.

6. The sense and antisense strands contain overhangs (blue color) for ligation into the pair of *Bbs*I sites in PX459 (CACC and CAAA, respectively). For transcription from the U6 promoter, add "G" nucleotide (red color) to the 5′end of the sense strand and its complementary "C" on the antisense strand.

7. Usually, there are no colonies on the negative control plate (no insert) because the *Bbs*I-digested plasmid alone cannot reanneal, and there are about hundreds of colonies on the sgRNA/pX459 cloning plate. Incomplete digestion of pX459 plasmid (uncut plasmid) can also result in colony formation. Before sequencing the plasmid, quick screening with restriction enzymes, *Bbs*I and *Eco*RI, can be performed. The LB plate can be stored at 4 °C (upside down) for up to 1 month.

8. The plasmids should be endotoxin-free and are prepared at a concentration greater than 2 μg/μL for efficient nucleofection. The Geneaid™ Midi Plasmid Kit produces endotoxin-free plasmids. Other kits with similar features can also be used.

9. Gently tap the Nucleocuvette™ strip to make sure that there is no air bubble in the well before nucleofection. Place the vessel with closed lid into the retainer of the 4D-Nucleofector™ X Unit with proper orientation. Do not leave the cells in the Nucleofection™ solution for extended periods. This will lead to reduced viability and transfection efficiency.

10. Supplementation of SMC4 to mTeSR™1 medium improves viability of iPSCs during clonal isolation. We found that the use of RevitaCell™ and StemFlex™ medium (Thermo Fisher Scientific) also helps increase viability and cloning efficiency for the feeder-free culture of iPSCs.
11. To obtain a single cell per well, the number of cells seeded into each well of 96-well plates can vary among the cell lines. The seeding density can range from 1 to 20 cells/well. The next day, identify the well that contains a single cell. If there are too many empty wells, increase the number of seeded cells per well. If there are too many wells containing multiple cells, decrease the number of seeded cells.

Acknowledgments

M.W. is supported by Chalermphrakiat Grant, Faculty of Medicine Siriraj Hospital, Mahidol University, Thailand.

References

1. Ruangvutilert P (2007) Thalassemia is a preventable genetic disease. Siriraj Med J 59:330–333
2. Ye L, Chang JC, Lin C, Sun X, Yu J, Kan YW (2009) Induced pluripotent stem cells offer new approach to therapy in thalassemia and sickle cell anemia and option in prenatal diagnosis in genetic diseases. Proc Natl Acad Sci U S A 106:9826–9830
3. Olivieri NF, Pakbaz Z, Vichinsky E (2011) Hb E/beta-thalassaemia: a common & clinically diverse disorder. Indian J Med Res 134:522–531
4. Papapetrou EP, Lee G, Malani N, Setty M, Riviere I, Tirunagari LMS, Kadota K, Roth SL, Giardina P, Viale A et al (2011) Genomic safe harbors permit high β-globin transgene expression in thalassemia induced pluripotent stem cells. Nat Biotechnol 29:73–81
5. Tubsuwan A, Abed S, Deichmann A, Kardel MD, Bartholoma C, Cheung A, Negre O, Kadri Z, Fucharoen S, von Kalle C et al (2013) Parallel assessment of globin lentiviral transfer in induced pluripotent stem cells and adult hematopoietic stem cells derived from the same transplanted beta-thalassemia patient. Stem Cells 31:1785–1794. https://doi.org/10.1002/stem.1436
6. Liu Y, Yang Y, Kang X, Lin B, Yu Q, Song B, Gao G, Chen Y, Sun X, Li X et al (2017) One-step biallelic and scarless correction of a beta-thalassemia mutation in patient-specific iPSCs without drug selection. Mol Ther Nucleic Acids 6:57–67. https://doi.org/10.1016/j.omtn.2016.11.010
7. Niu X, He W, Song B, Ou Z, Fan D, Chen Y, Fan Y, Sun X (2016) Combining single strand oligodeoxynucleotides and CRISPR/Cas9 to correct gene mutations in beta-thalassemia-induced pluripotent stem cells. J Biol Chem 291:16576–16585. https://doi.org/10.1074/jbc.M116.719237
8. Xie F, Ye L, Chang JC, Beyer AI, Wang J, Muench MO, Kan YW (2014) Seamless gene correction of beta-thalassemia mutations in patient-specific iPSCs using CRISPR/Cas9 and piggyBac. Genome Res 24:1526–1533. https://doi.org/10.1101/gr.173427.114
9. Song B, Fan Y, He W, Zhu D, Niu X, Wang D, Ou Z, Luo M, Sun X (2015) Improved hematopoietic differentiation efficiency of gene-corrected beta-thalassemia induced pluripotent stem cells by CRISPR/Cas9 system. Stem Cells Dev 24:1053–1065. https://doi.org/10.1089/scd.2014.0347
10. Wattanapanitch M, Damkham N, Potirat P, Trakarnsanga K, Janan M, Yaowalak UP, Kheolamai P, Klincumhom N, Issaragrisil S (2018) One-step genetic correction of hemoglobin E/beta-thalassemia patient-derived

iPSCs by the CRISPR/Cas9 system. Stem Cell Res Ther 9:46. https://doi.org/10.1186/s13287-018-0779-3

11. Ran FA, Hsu PD, Wright J, Agarwala V, Scott DA, Zhang F (2013) Genome engineering using the CRISPR-Cas9 system. Nat Protoc 8:2281–2308. https://doi.org/10.1038/nprot.2013.143

12. Byrne SM, Mali P, Church GM (2014) Genome editing in human stem cells. Methods Enzymol 546:119–138. https://doi.org/10.1016/B978-0-12-801185-0.00006-4

Chapter 15

CRISPR/Cas9-Mediated GFP Reporter Knock-in in K562 and Raji Cell Lines for Tracking Immune Cell Killing Assay

Nontaphat Thongsin and Methichit Wattanapanitch

Abstract

Cell-mediated cytotoxicity plays an important role in several fundamental immunological processes and is crucial for biological evaluation in in vitro studies. In order to determine the immunological activities of the cells, an assay should be safe, reproducible, and cost-effective. Here, we present a simple and cost-effective approach for evaluation of natural killer (NK) cell-mediated cytotoxicity by generating a CRISPR/Cas9-mediated GFP reporter knock-in in the target cell line, K562, and the non-target cell line, Raji, using a plasmid-based transfection method. The GFP^+ target cells facilitate tracking of the immune cell killing assay, which avoids the need for multiple cell labeling with fluorescent dyes. Our approach is also applicable to the genome editing of other target cell types for functional analysis of effector cells.

Key words CRISPR/Cas9, GFP knock-in, K562, Raji, Immune cell killing, NK-92 cells, Clonal selection

1 Introduction

Cell-mediated cytotoxicity is the most important effector pathway of immunological response against intracellular pathogens or tumor cells [1]. This process is considered a hallmark of $CD8^+$ cytotoxic T lymphocyte (CTL) and natural killer (NK) cell function [2, 3]. In addition, the mechanism is crucial for the biological evaluation of effector and/or target cell response in vitro [4]. There are two major contact-dependent cytotoxic pathways involved in the elimination of the target cells. The first pathway involves secretion of a pore-forming protein called perforin, and the pro-apoptotic serine protease family called granzyme B, which activates caspases. The second pathway involves the production of extrinsic apoptosis signaling proteins such as TNF-α, FasL, or TRAIL, which trigger apoptosis of the target cells [5, 6].

In the past decades, the radioactive chromium (^{51}Cr)-release assay was the first quantification method for cell-mediated cytotoxicity, as described by Brunner et al. [7]. This assay is based on the

Kumaran Narayanan (ed.), *Bio-Carrier Vectors: Methods and Protocols*, Methods in Molecular Biology, vol. 2211, https://doi.org/10.1007/978-1-0716-0943-9_15,

release of the radioactive probe from the pre-labeled lysed target cells to the culture medium. This assay is considered as the standard method for cell-mediated cytotoxicity and still widely used. However, high spontaneous release of labelled-^{51}Cr from the target cells and risks of radioisotope handling are the main drawback of this assay [8].

In order to avoid difficulties in using the radioisotope, enzyme-link immunospot assay (ELISpot) and flow cytometric assay have been developed. Cell-mediated cytotoxicity using ELISpot is based on the quantification of locally secreted cytokines captured by antibody-coated vessels. ELISpot provides more sensitivity and specificity than the ^{51}Cr release assay and circumvents the issues that occurred during cell labeling and spontaneous ^{51}Cr leakage. However, ELISpot does not provide a direct measurement of the target cell lysis and cytotoxicity mediated by FasL pathway [9, 10].

To overcome these limitations, numerous flow cytometric approaches have been applied for insight measurement of effector cells mediated target cell killing. These approaches include evaluation of caspase activity, annexin V binding apoptotic cells, uptake of propidium iodide (PI), or 7-amino-actinomycin D (7-AAD) by dying target cells [9]. However, to differentiate between the effector and target cell populations, additional cell labeling with fluorescent dyes such as carboxyfluorescein succinimidyl ester (CFSE) [11, 12] or the fluorescent protein such as green fluorescence protein (GFP) fusion tag [13] is required to facilitate the determination of effector and target cell populations.

Here, we use CRISPR/Cas9 genome editing technique to generate target cells, which stably express GFP. CRISPR/Cas9 system is a recent genome editing tool, which can precisely modify the target DNA sequence. The CRISPR/Cas9 component is composed of Cas9 protein, a CRISPR RNA (crRNA) of 20 nucleotides, which binds to the complementary target DNA, and a non-coding trans-activating CRISPR RNA (tracrRNA). The crRNA and tracrRNA can be fused together to form a single guide RNA (sgRNA) [14].

Interaction of sgRNA and target DNA facilitates the activation of the catalytic domain of Cas9 nuclease and results in a double-stranded break (DSB) at three nucleotides upstream of the 5′-NGG-3′ (PAM) site [15–17]. In response to DSB, the cells employ two endogenous repair mechanisms, either non-homologous end joining (NHEJ) or homology-directed repair (HDR) pathway. The NHEJ pathway repairs the DNA damage by ligating the broken ends of the DNA strand, independent of sequence homology. This repair mechanism is, therefore, error-prone, causing base insertions or deletions (indels) at the lesion site. In contrast, the HDR pathway employs the existing DNA molecule with the sequence homology to the region around the DSB as a template to repair the DSB. This mechanism can be exploited to generate precise gene modifications [16, 18].

In this chapter, we describe the protocol for precise GFP gene knock-in in K562 and Raji cells at a genomic safe harbor locus (*AAVS1*) using a plasmid-based CRISPR/Cas9 technology. Upon transfection with the CRISPR/Cas9 and the GFP donor plasmids, both of the transfected cells express high levels of GFP. Further enrichment using fluorescence-activated cell sorting (FACS) or limiting dilution generates a homogeneous population of GFP$^+$ cells, which can be directly used for NK cell killing assay. Our approach is easily performed and applicable to the GFP knock-in in various target cell types for functional analysis of effector cells.

2 Materials

2.1 Culture of K562, Raji, and NK-92 Cell Line

1. K562 and Raji cell culture medium: RPMI-1640, 10% FBS, 2 mM GlutaMAX™, 100 U/mL Penicillin-Streptomycin.
2. K562 and Raji cell cryopreservation medium: 90% FBS and 10% DMSO.
3. NK-92 cell culture medium: α-MEM, 10% FBS, 10% horse serum, 2 mM GlutaMAX™, 100 U/mL Penicillin-Streptomycin, 100 U/mL hIL-2.
4. NK-92 cell cryopreservation medium: 50% FBS, 40% NK-92 cell culture medium, and 10% DMSO.
5. 70% ethanol.
6. 15-mL conical tubes.
7. 25-cm^2 cell culture flasks.
8. Slow cooling cryo-container.
9. Cryogenic vials.
10. Hemocytometer.
11. Sterile seropipettes, pipette tips (200 and 1000 μL).
12. Tissue culture hood and humidified incubator (37 °C, 5% CO_2).
13. Inverted microscope.
14. Liquid nitrogen tank.

2.2 Preparation of Plasmids

1. Chemically competent DH5α *E. coli* cells.
2. gRNA_AAVS1-T2 (Addgene): sgRNA plasmid.
3. Plasmid hCas9 (Addgene): Cas9 plasmid.
4. AAV-CAG-EGFP (Addgene): GFP donor plasmid.
5. Glycerol stock of DH5α *E. coli* transformed with sgRNA plasmid, Cas9 plasmid, and GFP donor plasmid.

6. LB-ampicillin and LB-kanamycin: 25 g/L of Difco™ LB broth and 100 μg/mL ampicillin or 50 μg/mL kanamycin.
7. LB-ampicillin agar and LB-kanamycin agar: 35 g/L of LB agar, Lennox and 100 μg/mL ampicillin, or 50 μg/mL kanamycin.
8. 100 μg/mL Ampicillin.
9. 50 μg/mL Kanamycin.
10. 100% Glycerol.
11. 250-mL Erlenmeyer shake flasks.
12. Sterile pipette tips (2, 200, and 1000 μL).
13. 1.5-mL Eppendorf tubes.
14. Sterile cell spreader.
15. Tabletop centrifuge.
16. Incubator with shaker.
17. Plasmid mini- and midiprep kits.
18. 0.7% agarose gel: 0.7 g agarose powder dissolved in 100 mL 1× TBE buffer.
19. Novel juice (DNA staining reagent) (GeneDireX).
20. Standard ladder Lambda DNA/HindIII Marker (Thermo Fisher Scientific).
21. Standard ladder 1 Kb plus DNA ladder (Thermo Fisher Scientific).
22. Gel chamber and power source.
23. Gel documentation machine.
24. NanoDrop 8000 (Thermo Fisher Scientific).
25. Water bath.
26. Ice box.

2.3 Transfection of CRISPR/Cas9 and GFP Constructs into K562 and Raji Cell Lines

1. K562 and Raji cell culture medium: RPMI-1640, 10% FBS, 2 mM GlutaMAX™, 100 U/mL Penicillin-Streptomycin.
2. Purified sgRNA plasmid (gRNA_AAVS1-T2), Cas9 plasmid (hCas9), and GFP donor plasmid (AAV-CAG-EGFP).
3. P3 primary cell 4D-nucleofector X kit (Lonza).
4. 100 μL single Nucleocuvette™ (supplied with the kit).
5. Zombie Aqua™ Fixable Viability Kit (Biolegend).
6. 24-well plate.
7. 1.5-mL Eppendorf tubes.
8. Hemocytometer.
9. Sterile seropipettes, pipette tips (2, 200, and 1000 μL).
10. Inverted microscope.
11. Tabletop centrifuge.

12. Flow cytometer (LSRII).
13. Fluorescence microscope.
14. Amaxa™ 4D-Nucleofector™ (Lonza).
15. Tissue culture hood and humidified incubator (37 °C, 5% CO_2).

2.4 Enrichment of GFP+ Cells by Fluorescence-Activated Cell Sorting (FACS)

1. K562 and Raji cell culture medium: RPMI-1640, 10% FBS, 2 mM GlutaMAX™, 100 U/mL Penicillin-Streptomycin.
2. FAC sorting buffer: 2% FBS in 1× PBS.
3. 1% paraformaldehyde: 1 g of paraformaldehyde powder dissolved in 100 mL of 1× PBS.
4. Hemocytometer.
5. 15-mL conical tubes.
6. 5-mL round-bottom tubes with cap.
7. 5-mL round-bottom tubes with cell strainer cap.
8. Sterile seropipettes, pipette tips (2, 200, and 1000 μL).
9. Inverted microscope.
10. Fluorescence-activated cell sorter (FACSAria III).

2.5 Clonal isolation by Limiting Dilution

1. K562 and Raji cell culture medium: RPMI-1640, 10% FBS, 2 mM GlutaMAX™, 100 U/mL Penicillin-Streptomycin.
2. 96-well flat bottom plate.
3. Hemocytometer.
4. Multichannel pipette.
5. Sterile seropipettes, pipette tips (2, 200, and 1000 μL).
6. Fluorescence microscope.
7. Inverted microscope.
8. Tissue culture hood and humidified incubator (37 °C, 5% CO_2).

2.6 Immune Cell Killing

1. NK-92 cell line.
2. GFP^+ K562 cells.
3. GFP^+ Raji cells.
4. 96-well flat bottom plate.
5. Zombie Aqua™ Fixable Viability Kit (Biolegend).
6. Hemocytometer.
7. Sterile seropipettes, pipette tips (200 and 1000 μL).
8. Inverted microscope.
9. Flow cytometer (LSRII).
10. Tissue culture hood and humidified incubator (37 °C, 5% CO_2).

3 Methods

The following section describes a general protocol that we use for site-specific integration of the *GFP* gene in cell lines for tracking functional immune cell killing (Fig. 1). The protocol includes culture of the K562, Raji, and NK-92 cell lines, preparation of individual plasmids containing sgRNA, Cas9 and GFP genes, transfection of the plasmids into the cell lines, enrichment of the GFP^+ cells by FACS and limiting dilution. After clonal expansion, the engineered GFP^+ K562 or Raji cells can be used as target or non-target cells for the NK-92 cell killing, respectively. This approach facilitates immune cell killing assay since we can easily track and directly evaluate the efficiency of immune cell cytotoxic function.

3.1 Culture of K562 and Raji Cell Lines

3.1.1 Thawing

1. Warm K562 and Raji cell culture medium to room temperature.
2. Quickly thaw the cryopreserved K562 and Raji cells in a 37 °C water bath until only a small ice crystal remains.
3. Wipe the outside of the cryovial with 70% ethanol.
4. Gently pipette the cell suspension and transfer to a 15-mL conical tube containing 9 mL of RPMI-1640 basal medium.
5. Centrifuge the cells at $500 \times g$ for 5 min at room temperature.
6. Discard the supernatant and resuspend the pellet in 5 mL of culture medium.
7. Transfer the cell suspension into a 25-cm^2 cell culture flask and incubate at 37 °C, 5% CO_2 (*see* **Note 1**).

3.1.2 Passaging

Passaging is performed when the cells reach a maximum density: K562 cell line, 1×10^6 viable cells/mL, and Raji cell line, 3×10^6 viable cells/mL.

1. Gently resuspend the cells using a serological pipette and transfer the cell suspension to a 15-mL conical tube.
2. Aliquot a small amount of the cell suspension into a 1.5-mL Eppendorf tube for cell counting.
3. Centrifuge the cells at $500 \times g$ for 5 min at room temperature.
4. During centrifugation, perform cell counting using trypan blue exclusion assay to determine the cell number and viability.
5. After centrifugation, discard the supernatant and resuspend the pellet in 1 mL of culture medium.
6. Transfer the cells to a 25-cm^2 cell culture flask containing 5 mL of culture medium, adjust the seeding density to 1×10^5 viable cells/mL for the K562 cell line and 4×10^5 viable cells/mL for the Raji cell line.

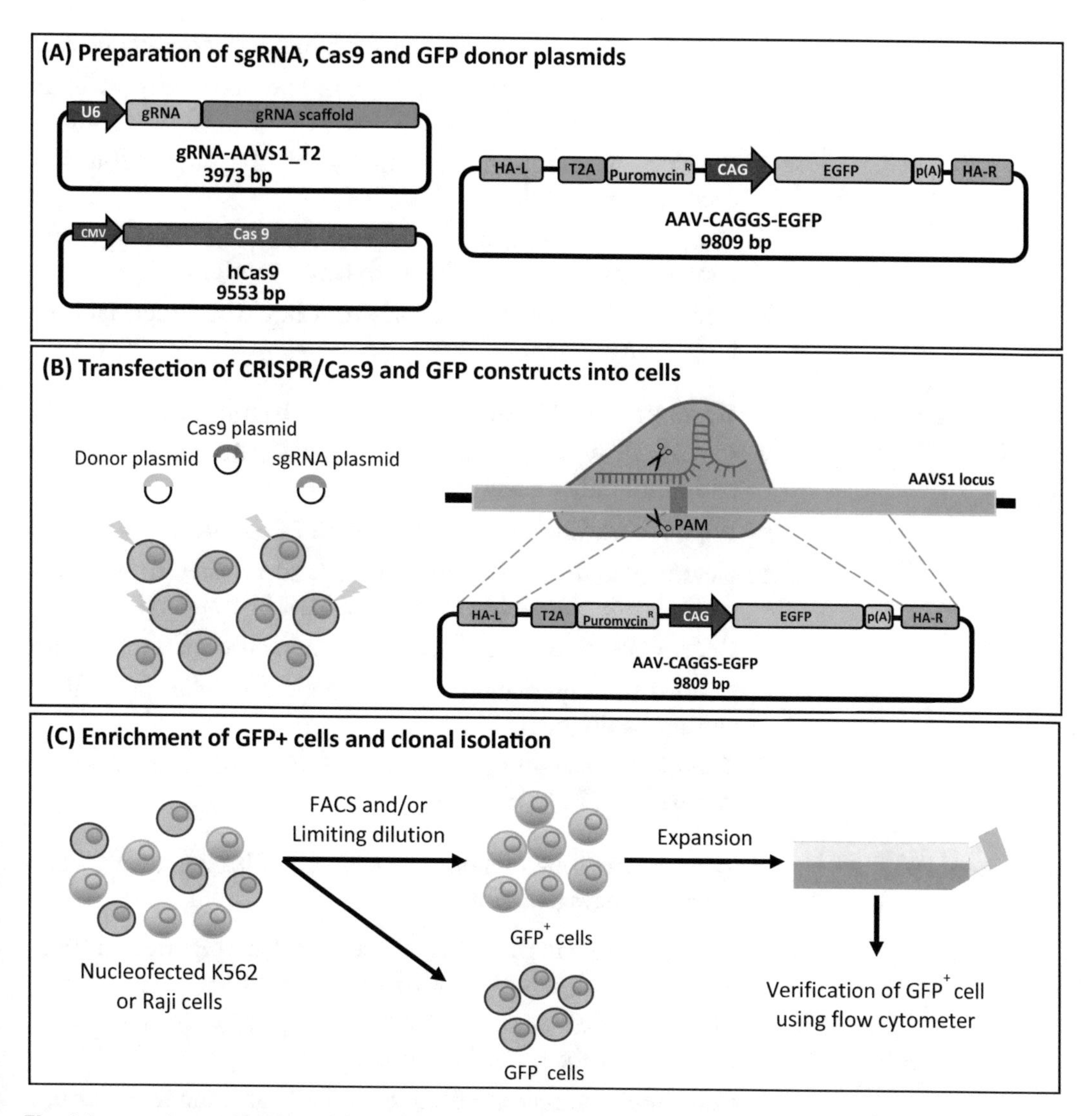

Fig. 1 An overview of CRISPR/Cas9-mediated genome editing for the generation of the GFP$^+$ K562 and Raji cell lines. (**a**) Schematic of genome editing plasmids including sgRNA plasmid (gRNA-AAVS1_T2), Cas9 plasmid (hCas9), and GFP donor plasmid (AAV-CAGGS-EGFP) harboring 5′ and 3′ homology arms of *AAVS1* gene. (**b**) Transfection of the three genome editing plasmids results in the incorporation of the GFP gene cassette at the *AAVS1* locus via homology-directed repair (HDR). (**c**) The GFP$^+$ cells are subsequently enriched using FACS or limiting dilution

7. Incubate at 37 °C, 5% CO_2.
8. Change the medium and adjust the cell density every 3–4 days.

3.1.3 Cryopreservation

1. Prepare K562 and Raji cell cryopreservation medium and chill at 4 °C.
2. Gently resuspend the cells and transfer the suspension into a 15-mL conical tube.

3. Centrifuge the cells at 500 × *g* for 5 min at room temperature.
4. Discard the supernatant and resuspend the pellet with the cryopreservation medium at a density of 2×10^6 cells/mL.
5. Quickly dispense 1 mL aliquots of the suspension into each cryogenic vial.
6. Transfer the cryogenic vials into the slow cooling cryo-container and store at −80 °C freezer overnight.
7. The next day, transfer the vials to a liquid nitrogen tank for long-term storage.

3.2 Culture of NK-92 Cell Line

3.2.1 Thawing

1. Warm the NK-92 cell culture medium to room temperature.
2. Quickly thaw the cryopreserved NK-92 cells in a 37 °C water bath until only a small ice crystal remains.
3. Wipe the outside of the cryovial with 70% ethanol.
4. Gently pipette the cell suspension and transfer to a 15-mL conical tube containing 9 mL of α-MEM basal medium.
5. Centrifuge the cells at 500 × *g* for 5 min at room temperature.
6. Discard the supernatant and resuspend the pellet in 5 mL of culture medium.
7. Transfer the cell suspension into a 25-cm^2 cell culture flask and incubate at 37 °C, 5% CO_2.

3.2.2 Passaging

Passaging is performed when the cells reach 80% confluence, usually every 3–4 days (*see* **Note 2**).

1. Gently resuspend the cells using a serological pipette and transfer the cell suspension to a 15-mL conical tube.
2. Aliquot a small amount of the cell suspension into a 1.5-mL Eppendorf tube for cell counting.
3. Centrifuge the cells at 500 × *g* for 5 min at room temperature.
4. During centrifugation, perform cell counting using trypan blue exclusion assay to determine the cell number and viability.
5. After centrifugation, discard the supernatant and resuspend the cell pellet with 1 mL of culture medium.
6. Transfer the cells to a 25-cm^2 cell culture flask containing 5 mL of culture medium at a seeding density of 2×10^5 viable cells/mL.
7. Incubate at 37 °C, 5% CO_2.
8. Change the medium and adjust the cell density every 3–4 days.

3.2.3 Cryopreservation

1. Prepare the NK-92 cell cryopreservation medium and chill at 4 °C.
2. Gently resuspend the cells and transfer the suspension into a 15-mL conical tube.

3. Centrifuge the cells at 500 × *g* for 5 min at room temperature.
4. Discard the supernatant and resuspend the pellet with cryopreservation medium at a density of 2×10^6 cells/mL.
5. Quickly dispense 1 mL aliquots of the cell suspension into each cryogenic vial.
6. Transfer the cryogenic vials into the slow cooling cryo-container and store at −80 °C freezer overnight.
7. The next day, transfer the vials to liquid nitrogen for long-term storage.

3.3 Preparation of Plasmids

We generate the GFP knock-in K562 and Raji cells using the plasmid-based CRISPR/Cas9 system consisting of three individual plasmids: (1) sgRNA plasmid (gRNA-AAVS1_T2) containing the sgRNA DNA fragment, which targets *AAVS1* locus, under the U6 promoter, (2) Cas9 plasmid (hCas9), and (3) GFP donor plasmid (AAV-CAGGS-EGFP) harboring the 5′ and 3′ homology arms of *AAVS1* gene (~800 bp each), the Puromycin resistant gene, the CAG promotor, the GFP expression cassette, and the terminator sequence (poly(A)) (Fig. 1a).

3.3.1 Transformation and Validation of Plasmids

1. Thaw chemically competent DH5α *E. coli* cells and plasmids slowly on ice.
2. Add 100 ng of a plasmid to 50 μL of the chemically competent DH5α *E. coli* cells in a 1.5-mL Eppendorf tube.
3. Carefully mix the competent cells with the plasmid using a pipette and incubate on ice for 10 min.
4. Perform heat shock in a water bath at 42 °C for 45 s.
5. Immediately incubate the mixture on ice for 5 min.
6. Add 900 μL of LB medium without ampicillin or kanamycin and incubate in the shaker at 37 °C, at 200 rpm for 1 h.
7. Centrifuge the cell suspension at 500 × *g* for 5 min and aspirate 500 μL of supernatant.
8. Gently resuspend the pellet and plate 100 μL of the transformed cells onto the LB agar plates containing 100 μg/mL ampicillin or 50 μg/mL kanamycin, and incubate the plates upside down for overnight at 37 °C.
9. The next day, pick 5 colonies, inoculate a single colony into 7 mL of LB medium containing 100 μg/mL ampicillin or 50 μg/mL kanamycin, and incubate at 37 °C at 200 rpm overnight.
10. Aliquot 5 mL of culture for plasmid extraction using plasmid miniprep kit and store 2 mL of culture at 4 °C for short-term storage.
11. Quantify the plasmid yield using NanoDrop 8000.

Table 1
Reaction setup for plasmid digestion

Components	Amount (μL) for 1 reaction
Plasmid (up to 1 μg)	2
Restriction enzyme(s)	1
10× buffer	2
Molecular grade water	Up to 20
Total volume	**20**

12. Check the plasmid integrity by gel electrophoresis. Load the mixture of the extracted plasmid (*see* **Note 3**) and Novel juice on 0.7% agarose gel using 100 V for 30 min (*see* **Note 4**).
13. Store the plasmid at −20 °C.
14. Verify the extracted plasmid using restriction enzyme digestion (*see* **Note 5**). Set up the reaction for plasmid digestion (Table 1).
15. Incubate the mixture in the water bath at 37 °C for 10 min (*see* **Note 6**).
16. Perform gel electrophoresis, load 5 μL of the reaction mix and 1 μL of Novel juice on 1% agarose gel using 100 V for 30 min along with 1Kb plus DNA ladder.
17. After validation of the plasmids, add 1 mL of culture (Subheading 3.3.1, **step 10**) into 100 mL of LB medium containing 100 μg/mL ampicillin or 50 μg/mL kanamycin and incubate in the shaker at 37 °C at 200 rpm overnight.
18. Perform plasmid extraction for 100 mL of bacterial culture using the midiprep kit and resuspend the plasmid with the endotoxin-free water.
19. Quantify the plasmid yield using NanoDrop 8000 and dilute the plasmid to 1000 ng/μL. Store the plasmid at −20 °C. This plasmid preparation will be used for nucleofection (*see* **Note 7**).

3.3.2 Cryopreservation

1. Inoculate the bacterial culture (Subheading 3.3.1, **step 10**) into 2 mL of LB medium containing 100 μg/mL ampicillin or 50 μg/mL kanamycin and incubate at 37 °C at 200 rpm overnight.
2. The following day, aliquot 500 μL of culture to 1.5-mL Eppendorf containing 500 μL of glycerol.
3. Mix thoroughly by pipetting and store at −80 °C freezer for long-term storage.

3.3.3 Thawing of DH5α *E. coli* Cell Harboring Plasmid

1. Remove an LB agar plate containing 100 μg/mL ampicillin or 50 μg/mL kanamycin from 4 °C refrigerator and warm the plates at room temperature.
2. Remove the DH5α *E. coli* cells harboring plasmid from −80 °C freezer (Subheading 3.3.2, **step 3**), open the tube and use a sterile loop to scrape off the frozen bacteria from the top.
3. Streak the bacteria onto the LB agar plate and incubate the plate upside down overnight at 37 °C.
4. The following day, pick the colony and inoculate into the LB medium containing antibiotic or store the culture plate at 4 °C.

3.4 Transfection of CRISPR/Cas9 and GFP Constructs into K562 and Raji Cell Lines

This section is adapted from a 4D-Nucleofector™ Protocol for K562 and Raji cells using 4D-Nucleofector™ X Unit. We perform the transfection in 100-μL Nucleocuvette™ Vessels (*see* **Note 8**) using the plasmids (Subheading 3.3).

3.4.1 Nucleofection

1. Change the medium for the K562 and Raji cells 24 h prior to transfection with plasmids for CRISPR/Cas9-mediated gene editing. Transfer the culture to a 15-mL conical tube and centrifuge at 500 × *g* for 5 min at room temperature. Discard the supernatant and resuspend the pellet in 5 mL of culture medium. Transfer the suspension to a 25-cm^2 cell culture flask.
2. The following day, prepare a 6-well culture plate by adding 1 mL of culture medium into a well and pre-incubate the plate at 37 °C, 5% CO_2.
3. Prepare 100 μL of P3 Nucleofector™ Solution by adding 18 μL of Supplement to 82 μL of Nucleofector™ Solution. Mix thoroughly by pipetting and chill the mixture on ice.
4. Harvest 1×10^6 K562 or Raji cells to a 15-mL conical tube and centrifuge at 500 × *g* for 5 min at room temperature.
5. Discard the supernatant and resuspend the pellet in the P3 Nucleofector™ Solution (*see* **Note 9**).
6. Add 1 μg of each plasmid to the cell suspension and mix thoroughly by pipetting (*see* **Note 10**).
7. Carefully transfer the mixture into the Nucleocuvette™ vessel using the supplied pipette.
8. Gently tap the Nucleocuvette™ vessel to ensure that the sample covers the bottom of the cuvette.
9. Place the Nucleocuvette™ vessel into the retainer of the 4D-Nucleofector™ X Unit.
10. Start Nucleofection™ process by pressing "Start" on the display of 4D-Nucleofector™ Core Unit. Use the program FF-120 for K562 cells and the program DS-104 for Raji cells.

11. After completion, carefully remove the Nucleocuvette™ vessel from the retainer.
12. Resuspend the cells with 400 μL of culture medium and mix the cells by gently pipetting up and down for two to three times using the supplied pipette.
13. Carefully transfer 500 μL of the cell suspension into the 6-well plate containing pre-incubated culture medium (**step 2**).
14. Incubate the cells at 37 °C, 5% CO_2.

3.4.2 Evaluation of GFP Knock-in Cells

The following day, the expression of GFP protein can be observed under a fluorescence microscope or determined using a flow cytometer. Transient GFP protein expression should be monitored due to the presence of non-integrated GFP donor plasmid in the transfected cells without the HDR (*see* **Note 11**).

1. Harvest 1×10^5 transfected K562 or Raji cells to a 5-mL round-bottom tube with a cap.
2. Centrifuge at 500 × *g* for 5 min at room temperature.
3. Discard the supernatant and resuspend the pellet with 50 μl of FAC buffer.
4. Add 0.1 μL of Zombie Aqua™ Fixable Viability Kit, vortex briefly, and incubate for 15 min in the dark.
5. Add 3 mL of FAC buffer, vortex briefly, and centrifuge at 500 × *g* for 5 min at room temperature.
6. Discard the supernatant and fix the cells using 500 μL of 1% paraformaldehyde, the stained cells can be stored at 4 °C for up to 2 weeks.
7. Determine the GFP-positive population using a flow cytometer (LSRII) and analyze using the FlowJo software. Exclude the dead cells, which stained positive for Zombie Aqua™.

3.5 Enrichment of the GFP+ K562 and Raji Cells Using FACS

To enrich the GFP^+ cell population of the transfected K562 and Raji cells, we use FACS to sort the GFP^+ cells. In some occasions, the GFP^- cells potentially contaminate the GFP^+ cells after FAC sorting. Further enrich the sorted cells using FACS or limiting dilution (Subheading 3.6) to obtain high purity of GFP^+ cells (Fig. 2a).

1. Harvest 4×10^6 transfected K562 or Raji cells to a 15-mL conical tube and centrifuge at 500 × *g* for 5 min at room temperature.
2. Discard the supernatant and resuspend the cell pellet with 1 mL of FACS buffer.
3. Filter the cell suspension using a 5-mL round-bottom tube with a strainer cap.

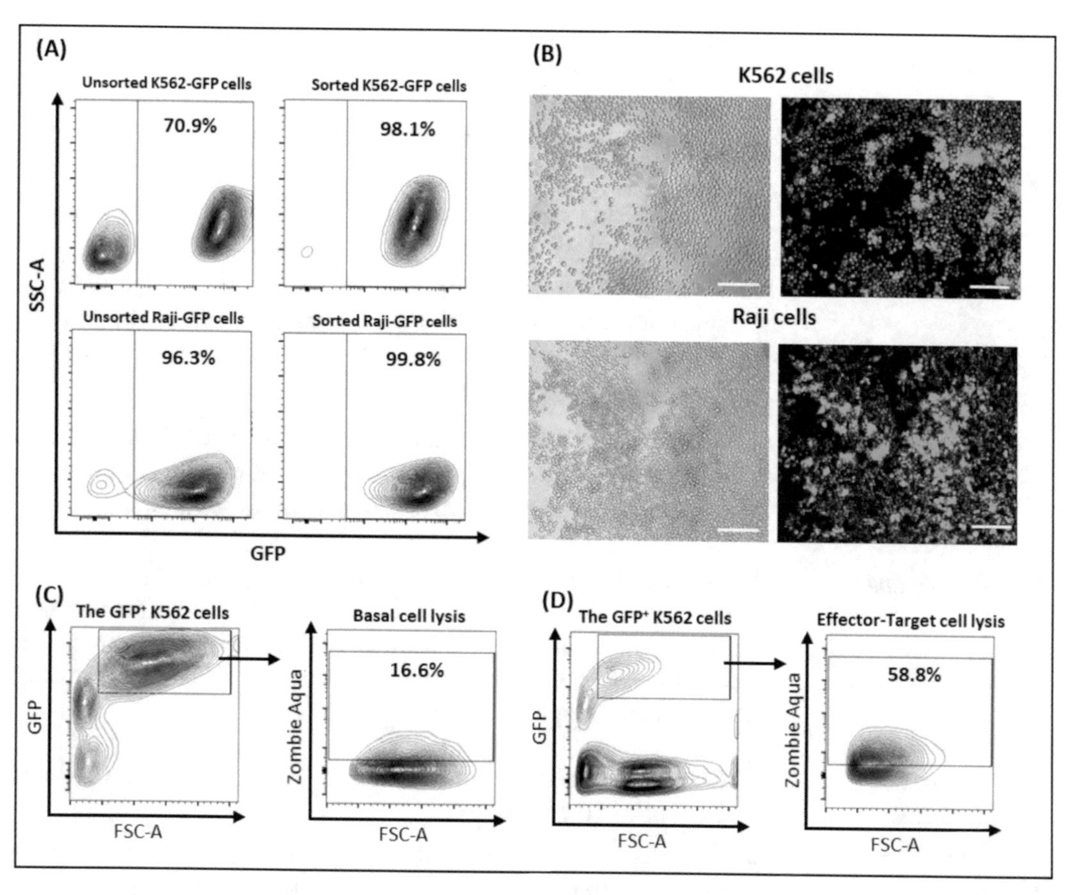

Fig. 2 Evaluation of the GFP$^+$ K562 and Raji cells and tracking of the target cells for cytotoxicity analysis. (**a**) Determination of the GFP$^+$ cells after limiting dilution in the unsorted and FAC sorted K562 and Raji cells using flow cytometry. (**b**) Expression of the GFP in K562 and Raji cells under the fluorescence microscope. Scale bar = 200 μm. (**c**, **d**) Determination of NK-92 cell killing capacity using the GFP$^+$ K562 cells as target cells by Flow cytometer. The gating strategy shows the GFP$^+$ K562 target cells after 5 h of co-culture with the NK-92 cell line at 5:1 (E:T) ratio. The dead GFP$^+$ K562 cells are positive for Zombie Aqua™. Basal target cell culture and effector-target cell culture are shown in panels (**c**) and (**d**), respectively

4. Chill the cells on ice and proceed with FACS immediately.
5. Use the non-transfected K562 or Raji cells as a gating control for FACS.
6. Sort the GFP$^+$ K562 or Raji cells into a 5-mL round-bottom tube with a cap containing 500 μL of culture medium.
7. Centrifuge at 500 × *g* for 5 min at room temperature.
8. Discard the supernatant, resuspend the pellet with the culture medium, and transfer the sorted cells into a culture vessel (*see* **Note 12**).
9. Incubate the cells at 37 °C, 5% CO_2.
10. On day 3 post-FAC sorting, transfer the cells to a 25-cm^2 cell culture flask.

11. On day 6 of culture, harvest the cells to evaluate the GFP$^+$ population using a flow cytometer (LSRII).

3.6 Clonal Isolation by Limiting Dilution

1. Prepare a 96-well flat-bottom plate containing 100 μL of culture medium/well.
2. Transfer the GFP sorted cells to a 15-mL conical tube and centrifuge at 500 × *g* for 5 min at room temperature.
3. Discard the supernatant and resuspend the pellet with an appropriate volume of culture medium to obtain the seeding density of 5 cells/well.
4. Incubate the cells at 37 °C, 5% CO_2.
5. After 2 weeks, transfer the GFP$^+$ cells from the 96-well plate to a larger well with the appropriate seeding density (Fig. 2b).
6. Evaluate the GFP$^+$ population using a flow cytometer (LSRII).

3.7 Immune Cell Killing

To evaluate the cytotoxic activity of the NK-92 cell line (effector cells), we co-culture the NK-92 cells with the GFP$^+$ K562 or Raji cells at 5:1 (effector: target) ratio for 5 h. The cytotoxic activity can be determined by tracking the GFP$^+$/Zombie Aqua™$^+$ cells using flow cytometer (Fig. 2c, d).

3.7.1 Co-culture of the NK-92 Cells with the GFP$^+$ K562 or Raji Cells

1. Change the medium for the NK-92, GFP$^+$ K562, and Raji cells 24 h prior to co-culture.
2. The next day, perform cell counting using trypan blue exclusion assay to determine the cell number and viability.
3. Aliquot 3 × 10^5 of the NK-92 cells and 6 × 10^4 of the GFP$^+$ K562 or Raji cells into 15-mL tubes.
4. Centrifuge at 500 × *g* for 5 min at room temperature.
5. Discard the supernatant and resuspend the pellet using 50 μL of complete RPMI culture medium and transfer the cell suspension into 96-well plate.
6. Incubate the cells at 37 °C, 5% CO_2 for 5 h.
7. Transfer the cells to a 5-mL round-bottom tube with a cap.
8. Centrifuge at 500 × g for 5 min at room temperature.
9. Resuspend the pellet with 50 μL of FACS buffer.
10. Add 0.1 μL of Zombie Aqua™ Fixable Viability Kit, vortex briefly, and incubate for 15 min in the dark.
11. Add 3 mL of FAC buffer, vortex briefly, and centrifuge at 500 × *g* for 5 min at room temperature.
12. Discard the supernatant and resuspend the cells using 500 μL of FAC buffer.
13. Immediately perform flow cytometry analysis.

3.7.2 Evaluation of Cytotoxic Activity Using Flow Cytometry Analysis

1. Determine the viability of the target cells using Zombie Aqua™ Fixable Viability Kit in the GFP^+ population.
2. Killing capacity of the NK-92 cells can be calculated using the following formula:

$$\%\text{Specific Lysis} = \frac{[100 \times (\%\text{Sample Lysis} - \text{Basal Lysis})]}{(100 - \%\text{Basal Lysis})}$$

The sample lysis represents the cell lysis in the presence of effectors at a given effector to target ratio whereas the basal lysis represents the cell lysis in the absence of effector cells.

4 Notes

1. After thawing, we recommend culturing the cells for at least 3 days for cell recovery and keep monitoring the cell density for passaging.
2. The NK-92 cells tend to grow in aggregates and are sensitive to overgrowth and medium exhaustion. Ensure that the cells are cultured in the appropriate cell density with regular medium change.
3. The amount of loaded plasmid for gel electrophoresis depends on the concentration of the extracted plasmid, at least 10 ng of the extracted plasmid should be used for qualification.
4. The percentage of agarose gel is dependent on the length of the DNA fragment. We use 0.7% agarose gel for evaluation of plasmid, genomic DNA, or large DNA fragment, and 1–1.5% agarose gel for the DNA fragment smaller than 1000 bp.
5. It is important to choose the appropriate restriction enzyme for plasmid digestion. We recommend using the restriction enzyme(s) that can cut at least 2 sites with different fragment lengths.
6. The reaction components and incubation time for plasmid digestion depend on the restriction enzyme; the digestion condition should follow the manufacturer's instruction. In this protocol, we use the FastDigest restriction enzyme (Thermo Fisher Scientific).
7. The plasmids used for transfection should be of high quality as poor quality of plasmid will decrease the transfection efficiency. We recommend concentrating the plasmid using the ethanol precipitation method. However, this step is not necessary if the amount of loaded plasmid for nucleofection does not exceed 10% of the total volume.
8. Alternatively, 20-μL Nucleocuvette™ Strip can be used. In this case, the number of the cells and the amount of plasmid should

be adjusted accordingly as recommended by the 4D-Nucleofector™ protocol.

9. It is important to work as quickly as possible. Leaving the cells in Nucleofector™ Solution for an extended period can lead to reduced cell viability and transfection efficiency.
10. The amount of the cells and plasmids for transfection can increase up to fivefold. However, the volume of loaded plasmid should not exceed 10% of the total volume.
11. To monitor the transient GFP expression in the transfected cells, we recommend performing flow cytometry analysis. The expression of GFP should be monitored for at least 2 weeks after transfection.
12. To obtain a proper growth rate after FACS, the cells should be seeded at the optimal density as recommended in an appropriate culture vessel.

Acknowledgments

We thank Pa-thai Yenchitsomanus for kindly providing us the K562 and Raji cell lines and Aussara Panya for the NK-92 cell line. This study was supported by a grant from Siriraj Research Fund, Faculty of Medicine Siriraj Hospital, Mahidol University (grant number (IO) R016333015 to MW). M.W. is supported by Chalermphrakiat Grant, Faculty of Medicine Siriraj Hospital, Mahidol University. N.T. is supported by Siriraj Graduate Scholarship, Faculty of Medicine Siriraj Hospital, Mahidol University, and the Royal Golden Jubilee Ph.D. Programme.

References

1. Lecoeur H, Février M, Garcia S, Rivière Y, Gougeon M-L (2001) A novel flow cytometric assay for quantitation and multiparametric characterization of cell-mediated cytotoxicity. J Immunol Methods 253(1):177–187. https://doi.org/10.1016/S0022-1759(01)00359-3
2. Uzhachenko RV, Shanker A (2019) CD8+ T lymphocyte and NK cell network: circuitry in the cytotoxic domain of. Immunity 10:1906. https://doi.org/10.3389/fimmu.2019.01906
3. Topham NJ, Hewitt EW (2009) Natural killer cell cytotoxicity: how do they pull the trigger? Immunology 128(1):7–15. https://doi.org/10.1111/j.1365-2567.2009.03123.x
4. Aslantürk ÖS (2018) In vitro cytotoxicity and cell viability assays: principles, advantages, and disadvantages. In: ML. Larramendy, & S. Soloneski (eds). Genotoxicity-A predictable risk to our actual world, vol 2. InTechOpen, London, UK. https://doi.org/10.5772/intechopen.71923
5. Lieberman J (2003) The ABCs of granule-mediated cytotoxicity: new weapons in the arsenal. Nat Rev Immunol 3(5):361–370. https://doi.org/10.1038/nri1083
6. Zhu Y, Huang B, Shi J (2016) Fas ligand and lytic granule differentially control cytotoxic dynamics of natural killer cell against cancer target. Oncotarget 7(30):47163–47172. https://doi.org/10.18632/oncotarget.9980
7. Brunner KT, Mauel J, Cerottini JC, Chapuis B (1968) Quantitative assay of the lytic action of immune lymphoid cells on 51-Cr-labelled allogeneic target cells in vitro; inhibition by isoantibody and by drugs. Immunology 14(2):181–196

8. Neri S, Mariani E, Meneghetti A, Cattini L, Facchini A (2001) Calcein-acetyoxymethyl cytotoxicity assay: standardization of a method allowing additional analyses on recovered effector cells and supernatants. Clin Diagn Lab Immunol 8(6):1131. https://doi.org/10.1128/CDLI.8.6.1131-1135.2001
9. Zaritskaya L, Shurin MR, Sayers TJ, Malyguine AM (2010) New flow cytometric assays for monitoring cell-mediated cytotoxicity. Expert Rev Vaccines 9(6):601–616. https://doi.org/10.1586/erv.10.49
10. Kalyuzhny AE (2005) Chemistry and biology of the ELISPOT assay. In: Kalyuzhny AE (ed) Handbook of ELISPOT: methods and protocols. Humana Press, Totowa, NJ, pp 15–31. https://doi.org/10.1385/1-59259-903-6:015
11. Kandarian F, Sunga GM, Arango-Saenz D, Rossetti M (2017) A flow cytometry-based cytotoxicity assay for the assessment of human NK cell activity. J Vis Exp 126:56191. https://doi.org/10.3791/56191
12. Rabinovich PM, Zhang J, Kerr SR, Cheng B-H, Komarovskaya M, Bersenev A, Hurwitz ME, Krause DS, Weissman SM, Katz SG (2019) A versatile flow-based assay for immunocyte-mediated cytotoxicity. J Immunol Methods 474:112668. https://doi.org/10.1016/j.jim.2019.112668
13. Csepregi R, Temesfői V, Poór M, Faust Z, Kőszegi T (2018) Green fluorescent protein-based viability assay in a multiparametric configuration. Molecules 23(7):1575. https://doi.org/10.3390/molecules23071575
14. Jinek M, Chylinski K, Fonfara I, Hauer M, Doudna JA, Charpentier E (2012) A programmable dual-RNA–guided DNA endonuclease in adaptive bacterial immunity. Science 337 (6096):816. https://doi.org/10.1126/science.1225829
15. Thurtle-Schmidt DM, Lo T-W (2018) Molecular biology at the cutting edge: a review on CRISPR/CAS9 gene editing for undergraduates. Biochem Mol Biol Educ 46(2):195–205. https://doi.org/10.1002/bmb.21108
16. Wang H, Russa ML, Qi LS (2016) CRISPR/Cas9 in Genome editing and beyond. Annu Rev Biochem 85(1):227–264. https://doi.org/10.1146/annurev-biochem-060815-014607
17. Ran FA, Hsu PD, Wright J, Agarwala V, Scott DA, Zhang F (2013) Genome engineering using the CRISPR-Cas9 system. Nat Protoc 8 (11):2281–2308. https://doi.org/10.1038/nprot.2013.143
18. Ceccaldi R, Rondinelli B, D'Andrea AD (2016) Repair pathway choices and consequences at the double-Strand break. Trends Cell Biol 26(1):52–64. https://doi.org/10.1016/j.tcb.2015.07.009

Part V

Analyzing and Profiling Gene Expression

Chapter 16

Shotgun Proteomics and Mass Spectrometry as a Tool for Protein Identification and Profiling of Bio-Carrier-Based Therapeutics on Human Cancer Cells

Syafiq Asnawi Zainal Abidin, Iekhsan Othman, and Rakesh Naidu

Abstract

Shotgun proteomics has been widely applied to study proteins in complex biological samples. Combination of high-performance liquid chromatography with mass spectrometry has allowed for comprehensive protein analysis with high resolution, sensitivity, and mass accuracy. Prior to mass spectrometry analysis, proteins are extracted from biological samples and subjected to in-solution trypsin digestion. The digested proteins are subjected for clean-up and injected into the liquid chromatography-mass spectrometry system for peptide mass identification. Protein identification is performed by analyzing the mass spectrometry data on a protein search engine software such as PEAKS studio loaded with protein database for the species of interest. Results such as protein score, protein coverage, number of peptides, and unique peptides identified will be obtained and can be used to determine proteins identified with high confidence. This method can be applied to understand the proteomic changes or profile brought by bio-carrier-based therapeutics in vitro. In this chapter, we describe methods in which proteins can be extracted for proteomic analysis using a shotgun approach. The chapter outlines important in vitro techniques and data analysis that can be applied to investigate the proteome dynamics.

Key words Shotgun proteomics, Protein identification, Protein profiling, Bio-carrier vectors

1 Introduction

1.1 Proteomics and Mass Spectrometry

Proteomics is defined as a large-scale analysis of proteome; a set of proteins produced in an organism, a system, or any biological context. The study of proteomics aimed to identify and quantify proteins in a proteome which encompasses the expression, localization, interactions, and post-translational modifications [1]. Proteomics has been described as an important approach to obtain biological information since most biological activities are attributed by proteins, thus enhancing our conception of a biological system [2]. Advancement of proteomics study has been driven by the

Kumaran Narayanan (ed.), *Bio-Carrier Vectors: Methods and Protocols*, Methods in Molecular Biology, vol. 2211, https://doi.org/10.1007/978-1-0716-0943-9_16, © Springer Science+Business Media, LLC, part of Springer Nature 2021

discovery of new technologies for protein separation, mass spectrometry technology, labeling techniques for protein quantification, and large-scale bioinformatics tool [1].

Mass spectrometry study has emerged as a multipurpose and comprehensive tool in analyzing large-scale proteomics due to its rapid advancement in resolution, sensitivity, mass accuracy, and protein analysis scan rate [1]. The system can distinguish different protein species by high precision measurements. The structural information is obtained from isolated and fragmented molecular ions using tandem mass spectrometry approach (MS/MS) [2].

Mass to charge ratio (m/z) of a molecule is applied to calculate the mass of a protein following an ionization process in the instrument. Ionization of proteins and peptides were achieved by methods such as electrospray ionization (ESI) and matrix-assisted laser desorption ionization (MALDI) [3, 4]. ESI produced ions from a high voltage electrospray to transform liquid analytes into aerosol while MALDI uses laser energy absorbing matrix to create ions from large molecules. ESI ion sources were originally applied on ion-trap or triple-quadrupole (QQQ) MS/MS, whereas MALDI was most often coupled with time-of-flight (TOF) analyzers. However, the availability of hybrid-quadrupole TOF (Q-TOF) MS/MS spectrometers have allowed and used frequently with ESI [2]. These instruments conduct isolation of specific ions based on their m/z ratio and the fragmentation of the ions within the gas phase and allow the MS/MS spectra to be recorded. The MS/MS spectrum of a peptide is then used as a basis to determine the amino acid sequence of a specific protein. Availability of sequence databases has greatly improved protein identification from mass spectrometric analysis by using algorithms that complement MS/MS spectra to the database [5, 6].

Protein characterization can be accomplished by different approaches such as the bottom-up or top-down proteomics. Bottom-up protein analysis or shotgun proteomics refers to the characterization of different proteins based on the analysis of peptides that were digested from the protein prior to analysis [1]. In this approach, proteins were typically digested with enzyme such as trypsin, fractionated, and subjected to LC-MS/MS analysis. The amino acid sequence and peptide identification can be achieved by comparing the MS/MS spectra of the digested peptides with theoretical MS/MS spectra from a protein database.

Unique or shared peptides of a particular protein will determine the confidence score and grouped according to their specific protein families [1]. In contrast, top-down proteomics is commonly applied for intact protein characterization. The method is advantageous in the determination of post-translational modifications and isoforms [7]; however, difficulties with protein fragmentation and ionization has made shotgun proteomics a much preferred method for protein analysis [1].

1.2 Bio-Carriers and Proteomics

The advancements of bio-carrier technology have been proven to be crucial in developing effective therapeutic agents against various diseases. Nanotechnology plays a key role in bio-carrier development by efficiently delivers drug to the target site with minimum side effects and improved drug reaction [8]. To date, various types of nano-sized bio-carrier (nanoparticles) and their therapeutic applications have been extensively reviewed [8–10]. In this chapter, the methodology to understand the proteomic changes in cell and/or tissue cultures upon interaction with bio-carrier-based drug is discussed.

2 Materials

2.1 Protein Samples, Extraction, and Quantification

1. Lyophilized protein samples (*see* **Note 1**).
2. Tissue homogenizer.
3. RIPA Lysis and Extraction Buffer (ThermoFisher, USA) with 1× Protease and Phosphatase inhibitor.
4. Pierce BCA Protein Assay Kit (ThermoFisher, USA).

2.2 Trypsin Digestion (In-Solution Protein Digestion)

1. Trypsin stock (20 μg of lyophilized trypsin, MS grade): Add 20 μL of 50 mM acetic acid to 20 μg of lyophilized trypsin.
2. 1.5 mL Protein Lo-bind centrifuge tubes (Eppendorf).
3. Trifluoroethanol (TFE).
4. *Digestion buffer*: 100 mM ammonium bicarbonate (ABC).
5. *Reducing buffer*: 200 mM dithiothreitol (DTT).
6. *Alkylation buffer*: 200 mM iodoacetamide (IAM).
7. Formic acid, MS grade.
8. Vacuum concentrator.

2.3 Peptide Concentration and Desalting

1. Pierce C18 Spin Column (ThermoFisher, USA).
2. 1.5 mL Protein Lo-bind centrifuge tubes.
3. Solutions (*see* **Note 2**).
   (a) *Activation solution*: 50% acetonitrile (ACN).
   (b) *Equilibration solution*: 0.5% formic acid (FA) in 5% ACN.
   (c) *Wash solution*: 0.5% FA in 5% ACN.
   (d) *Sample buffer*: 2% FA in 20% ACN (3 parts sample to 1-part sample buffer).
   (e) *Elution buffer*: 70% ACN.

2.4 Liquid Chromatography-Mass Spectrometry (LC-MS)

1. Agilent 1200 Series Nano and Capillary System LC with Agilent 6550 iFunnel Quadrupole-Time-of-Flight (Q-TOF) LC-MS, coupled with ChipCube nano-electrospray ionization (nano-ESI) source (*see* **Note 3**).

2. Buffer A: 0.1% FA in double distilled water (ddH_2O).
3. Buffer B: 0.1% FA in 90% ACN.

3 Methods

3.1 Protein Extraction and Quantification

1. Add approximately 200 μL of RIPA buffer (*see* **Note 4**) with 1× protease and phosphatase inhibitor to the sample and mix well (micropipette aspiration).
2. Homogenize the samples using 20 strokes of tissue homogenizer. Ensure that the sample remains on ice during the process.
3. Centrifuge at 500 × *g* for 10 min at 4 °C and collect the supernatant (dissolved, solubilized protein). The extracted protein lysate can be stored in −20 °C prior use or in −80 °C for long-term storage.
4. Perform protein quantitation using BCA assay with the Thermo Scientific Pierce BCA Protein Assay Kit (*see* **Note 5**).

3.2 Protein Identification: in-Solution Trypsin Digestion and Tandem Mass Spectrometry LC-MS/MS

3.2.1 In-solution Trypsin Digestion

1. Reconstitute protein samples in ddH_2O. Aliquot approximately 100 μg of proteins (estimated from the protein assay kit) in 100 μL from each sample and add into a 1.5 mL protein lo-bind centrifuge tube.
2. Mix 25 μL of 100 mM ABC, 25 μL of TFE, and 1.0 μL of DTT to the samples and vortex briefly (10 s). Heat the mixture for 20 min at 90 °C and cool down to room temperature (*see* **Note 6**).
3. Add 4 μL of IAM (alkylation step) to the sample and incubate at room temperature in the dark (avoid direct lighting/sunlight) for 1 h.
4. Add 1 μL of DTT to quench excess IAM and incubate at room temperature in the dark for 1 h.
5. Add 300 μL of ddH_2O and 100 μL of ABC to dilute the sample and add 1 μL of trypsin. Incubate the sample overnight at 37 °C.
6. Centrifuge the sample briefly (10 s), add 1 μL of formic acid to stop the enzymatic reaction, and concentrate the sample using a vacuum concentrator.

3.2.2 Peptide Clean-up/Desalting

1. Desalt and reconcentrate the digested peptides using Pierce C18 Spin Column (ThermoFisher, USA).
2. Prepare sample (digested proteins) by mixing 3 parts sample to 1 part sample buffer.
3. Tap the spin column gently to settle resin, remove the top and bottom cap of the spin column, and place the column into a receiver tube (1.5 mL microcentrifuge tube).

4. Add 200 μL of Activation Solution and centrifuge 1500 × *g* for 1 min. Discard the flow-through and repeat step.
5. Add 200 μL of Equilibration Solution and centrifuge 1500 × *g* for 1 min. Discard the flow-through and repeat step.
6. Load sample on top of the resin bed in the spin column and place the column into a receiver tube. Centrifuge 1500 × *g* for 1 min. Recover flow-through and reload on the resin bed to ensure complete sample binding.
7. Place column into a receiver tube and add 200 μL of Wash Solution and centrifuge at 1500 × *g* for 1 min. Discard flow-through and repeat step.
8. Place column into a clean receiver tube and add 20 μL of Elution Buffer on top of the resin bed. Centrifuge at 1500 × *g* for 1 min and repeat step using the same receiver tube.
9. Dry sample in a vacuum concentrator and suspend sample in 0.1% formic acid in water prior LC-MS/MS analysis.

3.2.3 Tandem Mass Spectrometry (LC-MS/MS)

1. Perform the protein identification using 1200 Series Nanoflow LC (Agilent, Santa Clara, CA, USA) connected to Accurate-Mass Q-TOF 6550 iFunnel with nano-electrospray ionization source.
2. Reconstitute the dried/lyophilized samples (digested peptides) in 10 μL of 0.1% formic acid in ddH$_2$O (Buffer A). Load the analytes to Large Capacity Chip LC Column, 300 Å, C18, 160 nL enrichment column with 75 μm × 150 mm analytical column (P/N: G4240–62010).
3. Set injection volume to 1 μL per sample and adjust the flow rate to 2 μL/min for capillary pump and 0.4 μL/min, with a linear gradient of 5–70% buffer B for 40 min.
4. Set mass spectrometer run parameters to 5.0 L/min at 325 °C for the drying gas, fragmentor voltage at 360 V and capillary voltage at 1800 V. Acquire the spectrum using Mass Hunter acquisition software (Agilent, CA, USA) in auto MS/MS mode with a mass range (m/z) of 110–3000 for MS and 50–3000 for MS/MS. Set the acquisition rate (spectra/s) for MS and MS/MS at 2 and 4, respectively.

3.3 Protein Data Analysis

This section will describe data mining using PEAKS Studio (Peptide de-novo sequencing, identification, and quantification software) (*see* **Note 7**).

1. Acquire the raw data from the mass spectrometer (.d format for Agilent instrument) and process with PEAKS Studio (Bioinformatic Solution Inc., Waterloo, Canada) version 7.5 against merged database (SwissProt and trEMBL) from Uniprot.org on the sample species origin.

2. Create a new project and select "mass only" as the correct precursor. Set the error tolerance in De Novo settings at 0.1 Da for both precursor and fragment mass. Select trypsin as the digestion enzyme and carbamidomethylation as the fixed post-translational modification (PTM). Set the maximum allowed variable PTM at 3 per peptide and report up to 5 candidates per spectrum.
3. In the database/PEAKS DB search, set the error tolerance for precursor mass at 0.1 Da using monoisotopic mass and fragment ion at 0.1 Da. Select trypsin once again as the digestion enzyme and allow nonspecific cleavage at both ends of the peptide. Set the maximum number of missed cleavages per peptide at 3 and carbamidomethylation as fixed PTM. Set the maximum number of allowed variable PTM pep peptide at 10.
4. Allow PEAKS software to perform additional analysis by selecting the following options:
   (a) Estimate FDR with decoy-fusion.
   (b) Find unspecified PTM and common mutations with Peaks PTM.
   (c) Find more mutations with SPIDER.
5. Validate the identified proteins with the following filters: Protein -10lgP score $\geq$ 20 or FDR 1% and containing at least two unique peptides for identification.
6. Export the data from the PEAKS Studio software to spreadsheet (Microsoft Excel) and tabulate the protein data based on the protein family (*see* **Note 8**).

4 Notes

1. Lyophilize protein samples and store at −20 °C until further use.
2. Solution: Prepare all solutions with double distilled water (ddH_2O).
3. Solutions for liquid chromatography: Sonicate all solutions for 20 min prior usage.
4. Protein extraction from tissues and/or cultured cells is the initial step in various biochemical techniques such as SDS-PAGE, western blot, mass spectrometry, and protein purification via liquid chromatography. Rapid and efficient protein extraction can be achieved by radioimmunoprecipitation assay (RIPA) buffer. The buffer can be used together with protease and phosphatase inhibitors to avoid protein degradation and maintain protein phosphorylation.

5. Normalized protein concentration is needed for sample comparison, and it can be achieved by protein quantitation assay. Bicinchoninic acid (BCA) assay offers significant advantage over Bradford assay for protein quantitation as it is extremely sensitive and compatible with most ionic and nonionic detergents. BCA assay works by the reduction of cupric ions by protein in alkaline conditions. Interaction of BCA with the cupric ions produces a colored (purple) water-soluble complex that can be detected using spectrophotometer at 562 nm.
6. Proteolytic digestion of proteins from heterogeneous lysate are able to circumvent several challenges related to mass spectrometry analysis on intact proteins such as the separation, ionization, and characterization [12]. Trypsin is a digestion enzyme that specifically cleaves at the carboxyl site of arginine and lysine. The tryptic digested peptides are typically analyzed by LC-MS/MS and allows for the identification and determination of protein expression levels through peptide abundance (fragment ion intensity) [5]. Typical in-solution trypsin digestion involves the reduction of peptide disulfide bonds by denaturing agent such as dithiothreitol (DTT). This is followed by alkylation process to prevent the re-formation of disulfide bonds by introducing a carbamidomethylation modification on the free cysteine residues. The alkylation step is important to allow maximum trypsin digestion efficiency.
7. Database searching through search engines are crucial in proteomic analysis as it can translate raw MS/MS spectra obtained into protein with identification data (protein ID). Examples of database search engines include Mascot, SEQUEST, PEAKS Studio, OMSSA, X!Tandem, and Agilent Spectrum Mill. In this chapter, the methodology is described based on PEAKS studio search engine as it offers more sensitive and accurate peptide identification [11].
8. The results table should display information such as the protein accession number (Protein ID), protein score (confidence, -10lgP score), protein coverage (%), number of peptides and unique peptides identified, average mass (MW), and protein description. This information is crucial to present as it highlights proteins that were detected with high confidence from the biological samples. Proteins identified with high confidence are generally determined by protein score of more than 20 and/or with unique peptides identified equal or more than 2.

References

1. Eng JK, McCormack AL, Yates JR (1994) An approach to correlate tandem mass spectral data of peptides with amino acid sequences in a protein database. J Am Soc Mass Spectrom 5 (11):976–989. https://doi.org/10.1016/1044-0305(94)80016-2
2. Fenn JB, Mann M, Meng CK, Wong SF, Whitehouse CM (1989) Electrospray ionization for mass spectrometry of large biomolecules. Science (New York, NY) 246 (4926):64–71
3. He H, Liang Q, Shin MC, Lee K, Gong J, Ye J, Liu Q, Wang J, Yang V (2013) Significance and strategies in developing delivery systems for bio-macromolecular drugs. Front Chem Sci Eng 7(4):496–507. https://doi.org/10.1007/s11705-013-1362-1
4. Karas M, Hillenkamp F (1988) Laser desorption ionization of proteins with molecular masses exceeding 10,000 daltons. Anal Chem 60(20):2299–2301
5. Leon IR, Schwammle V, Jensen ON, Sprenger RR (2013) Quantitative assessment of in-solution digestion efficiency identifies optimal protocols for unbiased protein analysis. Mol Cell Proteomics 12(10):2992–3005. https://doi.org/10.1074/mcp.M112.025585
6. Lombardo D, Kiselev MA, Caccamo MT (2019) Smart nanoparticles for drug delivery application: development of versatile nanocarrier platforms in biotechnology and nanomedicine. J Nanomater 2019:26. https://doi.org/10.1155/2019/3702518
7. Mann M, Kellner R, Lottspeich F, Meyer HE (2007) Sequence database searching by mass spectrometric data. In: Microcharacterization of Proteins. Wiley-VCH Verlag GmbH, pp 223–245. https://doi.org/10.1002/9783527615711.ch14
8. Patterson SD, Aebersold RH (2003) Proteomics: the first decade and beyond. Nat Genet 33:311
9. Saallah S, Lenggoro IW (2018) Nanoparticles carrying biological molecules: recent advances and applications. KONA Powder Particle J 35:89–111. https://doi.org/10.14356/kona.2018015
10. Tran JC, Zamdborg L, Ahlf DR, Lee JE, Catherman AD, Durbin KR, Tipton JD, Vellaichamy A, Kellie JF, Li M, Wu C, Sweet SM, Early BP, Siuti N, LeDuc RD, Compton PD, Thomas PM, Kelleher NL (2011) Mapping intact protein isoforms in discovery mode using top-down proteomics. Nature 480 (7376):254–258. https://doi.org/10.1038/nature10575
11. Zhang J, Xin L, Shan B, Chen W, Xie M, Yuen D, Zhang W, Zhang Z, Lajoie GA, Ma B (2012) PEAKS DB: de novo sequencing assisted database search for sensitive and accurate peptide identification. Mol Cell Proteomics 11(4):M111.010587. https://doi.org/10.1074/mcp.M111.010587
12. Zhang Y, Fonslow BR, Shan B, Baek M-C, Yates JR (2013) Protein analysis by shotgun/bottom-up proteomics. Chem Rev 113 (4):2343–2394. https://doi.org/10.1021/cr3003533

Index

Kumaran Narayanan (ed.), *Bio-Carrier Vectors: Methods and Protocols*, Methods in Molecular Biology, vol. 2211, https://doi.org/10.1007/978-1-0716-0943-9,

F

G

H

I

L

M

N

O

P

R

S

T

V

Z

Printed in the United States
by Baker & Taylor Publisher Services